Crossing the Digital Divide

Applying Technology to the Global Refugee Crisis

SHELLY CULBERTSON, JAMES DIMAROGONAS, KATHERINE COSTELLO, SERAFINA LANNA

Sponsored by Schmidt Futures

SOCIAL AND ECONOMIC WELL-BEING

For more information on this publication, visit www.rand.org/t/RR4322

Library of Congress Cataloging-in-Publication Data is available for this publication.
ISBN: 978-1-9774-0395-7

Published by the RAND Corporation, Santa Monica, Calif.
© Copyright 2019 RAND Corporation
RAND® is a registered trademark.

Support RAND
Make a tax-deductible charitable contribution at
www.rand.org/giving/contribute

www.rand.org

Preface

Approximately 71 million people globally are displaced by conflict and persecution, and this crisis has placed unprecedented strain on displaced individuals, host countries, and the international humanitarian system. Technology can mitigate some of these challenges. This report focuses on how technology is managed, used, perceived, and developed in refugee settings and on the ethical considerations for its use. This report should be of particular interest to foundations contributing financially to humanitarian situations, as well as to government donors, policymakers, practitioners, nongovernmental organizations, and private-sector companies engaged in helping refugees globally.

This research was sponsored by Schmidt Futures, which seeks to advance society through technology, inspire breakthroughs in scientific knowledge, and promote shared prosperity.

Community Health and Environmental Policy Program

RAND Social and Economic Well-Being is a division of the RAND Corporation that seeks to actively improve the health and social and economic well-being of populations and communities throughout the world. This research was conducted in the Community Health and Environmental Policy program within RAND Social and Economic Well-Being. The program focuses on such topics as infrastructure, science and technology, community design, community health promotion, migration and population dynamics, transportation, energy, and climate and the environment, as well as other policy concerns that are influenced by the natural and built environment, technology, and community organizations and institutions that affect well-being. For more information, email chep@rand.org.

Contents

Preface iii

Figures and Tables vii

Summary ix

Acknowledgments xix

Abbreviations xxi

CHAPTER ONE

Introduction 1

Study Approach 3

Limitations 5

Roadmap for This Report 6

CHAPTER TWO

**Roles and Responsibilities of Entities Involved in Using or Contributing to
    Technology in Refugee Settings** 7

Refugees 7

Aid Agencies 8

Host Countries 11

Donors 12

Technology Companies 13

Consortia 15

Universities and Research Organizations 17

Conclusion 17

CHAPTER THREE

Uses of Technology in Refugee Settings 19

Internet Connectivity and Access 19

Communication with Family and Friends 20

Information for Journeys 21

Establishment in New Locations 21

Language 23

Memory and Record Preservation. 24
Employment 24
Education 25
Aid Agency Management and Coordination 27
Distribution of Assistance. 27
Data Collection and Analysis. 29
Registration 31
Identity Management and Digital Identity Provision 32
Conclusion. 33

CHAPTER FOUR
Refugees' Perspectives on Technology. 35
Access to Technology. 35
Uses of Technology 39
Concerns About Technology 47
Conclusion. 49

CHAPTER FIVE
Business Models for Developing and Deploying Technology in Refugee Settings. 51
What Is a Business Model?. 52
Applying the Business Model Components to Developing and Deploying Technology
   in Refugee Settings 52
Barriers to (and Facilitators of) Developing and Deploying Technology in Refugee
   Settings. 58
Tools for Applying System-Level Thinking to Support the Development and
   Deployment of Technology in Refugee Contexts 62
Conclusion. 65

CHAPTER SIX
**Ethical, Security, and Privacy Issues Related to the Use of Technology in
   Refugee Settings** 67
Ethical Frameworks and Safeguards to Address Risks 67
Data Responsibility. 70
Bias. 74
Conflicts of Interest. 76
Conclusion. 77

CHAPTER SEVEN
Conclusions and Recommendations 79
Looking Ahead. 87

References. 89

Figures and Tables

Figures

S.1. Five Drivers for Evaluating the Application of a Technology Solution
      in Refugee Settings xv
5.1. Key Processes for Developing and Deploying a Technology Solution in
      Refugee Settings 54
5.2. Five Drivers for Evaluating the Application of a Technology Solution in
      Refugee Settings 64
7.1. A Business Model Canvas for Technology Investment in Refugee Settings 80

Tables

1.1. Distribution of the Focus Groups with Refugees and Internally Displaced
      Persons 4
5.1. Components of a Business Model 52
5.2. Questions to Ask to Guide a System-Level Approach to Developing and
      Deploying Technology in Refugee Settings 62

Summary

In the past two decades, the global population of forcibly displaced people has more than doubled, from 34 million in 1997 to 71 million in 2018. Of those displaced, 85 percent live in developing countries, many of which struggle to provide opportunities even for their own populations. In 2018, the United Nations ratified a new Global Compact on Refugees, which laid out new principles for supporting the self-sufficiency of refugees and emphasizing assistance for host countries so they can include refugees in their public services.

These principles and an overall momentum for change offer opportunities for new approaches to managing refugee situations, such as enhanced use of technology to solve problems in humanitarian settings. Technology has shown promise in many ways, such as facilitating the operations of aid organizations and providing a means of communication among displaced people. Yet there is room to expand and improve technology's use and effectiveness.

With this study, we aim to provide a systematic analysis of technology uses, needs, gaps, and opportunities for helping displaced people and responding agencies. Our primary focus is on digital technologies (often referred to as information and communication technologies) and systems that can be enhanced by digital technologies. We first seek a better understanding of the role of technology in refugee settings, so we look at the range of entities involved, the ways in which technology is used, and the perspectives of refugees on the uses and needs for technology. We then discuss business models to support the development and use of technology in refugee settings, as well as the ethical, privacy, and security issues associated with the use of technology in these settings. We close by offering recommendations to improve roles and responsibilities, streamline business processes, and address ethical and security concerns.

When conducting the study, we used multiple qualitative methods: a literature review of nearly 200 documents; 30 semi-structured interviews with stakeholders, including United Nations humanitarian agencies, implementing nongovernmental organization (NGO) partners, refugee leadership groups, private-sector technology companies, and government officials; and focus groups with refugees, internally displaced persons, or both in Bogotá and Cúcuta, Colombia; Athens, Greece; Amman, Jordan; Pittsburgh, Pennsylvania, United States; and the Maheba Refugee Camp in Solwezi, Zambia.

Roles and Responsibilities

There are many entities involved in using or contributing to technology in refugee settings, and refugees engage with technology in multiple ways. For example, refugees use technology, contribute to its development, and provide input to inform aid organizations about refugee needs. Multilateral organizations (such as the United Nations High Commissioner for Refugees and related agencies) and implementing partner NGOs draw on technology when managing overall refugee responses and programs for essential services. Host governments use technology in offering essential services to refugees and setting national legal frameworks governing the use of technology and data. Donor governments, multilateral organizations, aid agencies, and foundations use technology to provide funding, create priorities, track the use of funding and project implementation, and provide diplomatic leadership toward solutions. Technology companies provide both products that are typical of other circumstances and platforms specialized for refugee settings, and they donate skills and expertise. Consortia bring together different types of stakeholders to solve problems using technology, and universities provide research and innovation.

With all these entities that are engaged in some way with technology in response to the refugee crisis, changes in technology are resulting in changes to each entity's roles and responsibilities—for example, by creating new roles, simplifying long-standing roles, and altering the way that operations and programs are managed. At the same time, some of the people we interviewed perceived that more could be done to integrate technology in aid organizations' and other entities' operations by addressing specific technology needs, improving business processes, and developing legal and ethical considerations—all issues examined in this research.

Uses of Technology

We identified multiple ways in which technology can be used in the refugee context—for example, providing internet connectivity and access, supporting communication with family and friends, providing education and employment opportunities, facilitating distribution of housing and other resources, and providing a record of information about a displaced person's identity.

Refugees place a high value on internet and mobile connectivity, which allows them to keep in touch with family and friends, maintain documentation regarding their identity and experiences, and access information and support. Although many uses of technology have been developed specifically for refugee situations, most refugees typically rely instead on mainstream platforms (such as Facebook, WhatsApp, and YouTube). A criticism of the sizable amount of technology developed specifically for refugees is that much of it has been launched but not maintained, leading to the prevalence of *digital litter*—a trail of outdated information; broken links; and false impressions of rich, available digital tools.

Aid providers use various technologies to coordinate and manage their activities in supporting refugees, communicate about the availability of assistance, and distribute assistance. They use digital technologies to collect and analyze field and program data, registration data, survey responses, location information, and qualitative data, as well as other information from refugees, partner organizations, and publicly available sources. With the increasing reliance on data in aid operations, there is a need for humanitarian organizations to improve their technical capacity for managing and securing such data. Also of note is the growing use of cash assistance for aid distribution, which can rely on such technologies as blockchain (for recording transactions), biometrics (for identity), and credit card systems. A key gap is the need for refugees to have accepted proofs of identity for such needs as banking, purchasing, employment, and registration for services. Technological solutions present an opportunity to help fill this gap, although policies that govern digital identity remain undeveloped.

Refugee Perspectives

Our focus groups with refugees in the locations noted earlier provided individual perspectives on the multiple ways that refugees use technology.

Access to Technology

Refugees in each of the host countries where we conducted focus groups reported that they typically used mobile devices, primarily smartphones, and preferred these over computers or tablets because they were cheaper and easier to use while on the move. However, some refugees felt that the lack of access to a computer limited protection of their privacy and restricted their access to a wider range of programs. The majority of refugees stated that common platforms, such as Facebook and WhatsApp, were the most important and most frequently used because of their ubiquity, their community groups for refugees, and their low cost.

Refugees placed a high value on their ability to access the internet, although the vast majority of refugees across all six host countries described limited or irregular access to Wi-Fi and cellular data. Barriers to access included the cost of data plans and hardware, problems accessing SIM cards (subscriber identification modules) for smartphones, a lack of Wi-Fi where refugees were journeying or living, limited understanding of how to use technology, and a lack of language skills.

Uses of Technology

Refugees in the focus groups reported diverse uses of technology, which we grouped into the following categories:

- *Communication.* The vast majority of refugees across all six host countries stated that they most frequently used digital technology for communication with family, friends, and others. Refugees also expressed a sense of duty to share information—such as about access to services and assistance—online with a wider audience beyond family and friends.
- *Information for journeys and settling into new locations.* When internet connectivity was available, refugees used technology, especially transportation-related technology, to assist with their journeys from their home countries and to settle into their new locations. Refugees in Greece explicitly commented on the role that technology played when they hired smugglers to take them to Europe.
- *Language.* In ten of the 12 focus groups, participants mentioned using digital tools to learn English, while others used online or telephone translation services.
- *Education.* Both adults and children used digital technology to support their education. Adults did so to learn skills and information and to obtain certifications, and children used it to keep up with schooling, especially when their refugee status posed a barrier to formal education.
- *Employment.* Refugees used digital technology to seek employment opportunities, follow job trends, keep up to date with skills important to their careers, and pursue self-employment and entrepreneurship.
- *Faith-based activity.* Several refugees discussed using digital technology for religious purposes, including accessing the Quran and assistance opportunities at local churches.
- *Health care.* Refugees also used digital technology in seeking health care outside of traditional or official networks. This included using smartphone applications (apps) to find a doctor or to gain information to support self-care.
- *Identity management.* Many refugees described using digital technology, particularly their smartphones and cloud-based resources, to save, share, and acquire documents related to their personal identities and educational or professional qualifications.
- *Money management.* Some refugees had access to electronic money management, and some did not. Even where such services were available, refugees across almost all the countries where we conducted focus groups described money management as challenging because of legal restrictions on opening a bank account and lack of credit or debit cards.

Concerns About Technology

Many of the online security and privacy concerns described by refugees were similar to those of the general public (such as scams and improper use of personal data), although some issues arose specifically because they were refugees. In particular, refugees felt more vulnerable to fraud or data security breaches because of their refugee status and their unfamiliarity with the culture or language in their host country. In addition, refugees expressed mixed feelings about the presence of technology in their lives, including the costs of technology and its potential to turn into an addiction.

Business Models for Developing and Deploying Technology in Refugee Settings

We examined the business models through which technology is developed and implemented in support of refugees, and we considered how these models might be better applied when deploying such technology. A business model explains how and why the business creates value. In the private sector, the driving force of a business model is profit, and customers ultimately decide what works and what is relevant. In contrast, when it comes to deploying technology in support of refugees, recipients of technologies have less of a voice, and technology development is heavily influenced by United Nations agencies, governments, and NGOs. Nonetheless, certain elements of the traditional business model—identifying an opportunity; securing funding to develop the concept; demonstrating the concept; and deploying, scaling, and sustaining it—can apply in this context.

There are four main components of a business model as applied in the refugee context:

- The *value proposition* starts with the service or the need to be met, for either refugee populations or aid agencies. Value is derived not just from the technology itself but from but the way the service is provided and the end user's experience. Value can derive from a technology's ability to meet a critical need and lower costs, as well as from such features as flexibility, familiarity, accountability, and privacy.
- *Key resources* are any elements that can be used to create value, such as a network of relationships, an organization's image and level of trust, and internet connectivity.
- *Key processes* refer to the activities that need to take place for a product to be delivered to its intended users. We delineated six steps that occur in the typical development and implementation of a technology: (1) project initiation and con-

cept development, (2) product development and deployment, (3) content development, (4) training, (5) system sustainment and maintenance, and (6) system phaseout and retirement.

- The *profit formula* shows investors that the benefits are worth the costs. Benefits may be either direct (e.g., value created for the refugees or aid organizations) or indirect (e.g., development of intellectual property to be used in future business initiatives). Costs follow, for the most part, from the activities and processes associated with the product life cycle.

Building a More Systematic Approach to Technology Deployment

Several barriers and facilitators can influence whether the deployment of technology in refugee settings is successful. Barriers include a short-term rather than long-term mindset, a funding-driven rather than need-driven project, an emphasis on growth rather than on economies of scale, a focus on the technology rather than on changing business processes, and regulatory and organizational complexities. Furthermore, an overarching barrier is the lack of a system-level approach to thinking about technology in the context of refugee and humanitarian aid. This lack of clear system-level planning and execution makes it difficult for private-sector participants and NGOs to understand how they fit into the broader system and develop a business model that is viable and provides value.

Tools are available to build a more systematic approach to technology deployment in refugee contexts. As part of this study, we sketched out a broad system-level approach for evaluating the application of technology in a refugee setting (Figure S.1). Our approach focuses on the refugees as a vulnerable population and assesses five drivers that influence how effective the application will be.[1] We selected the five drivers based on interviewee comments. The region of origin, the host country, and the internal social and cultural pressures set the population's sociopolitical context and associated risks and constraints for applying any technology. The existing solutions and the complementary and contrasting activities describe the competitive and collaborative environment that could translate to risks and opportunities.

---

[1]   The approach is influenced by Michael Porter's five forces governing competition (Porter, 1979), except our focus is on the refugees instead of on the market.

Figure S.1
Five Drivers for Evaluating the Application of a Technology Solution in Refugee Settings

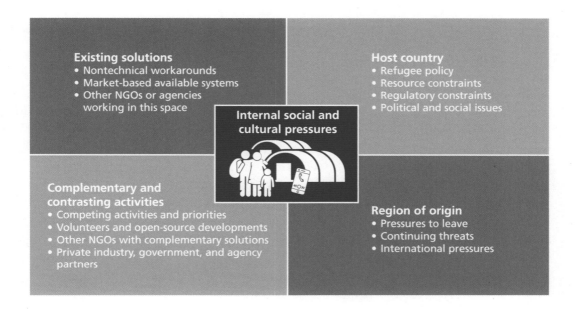

Ethical, Security, and Privacy Issues

Although technology has significant benefits for refugees and aid organizations, increased use of data raises ethical, security, and privacy issues. We identified four areas of consideration:

- *Frameworks and safeguards to address technology risks* are underdeveloped and fragmented across the humanitarian sector, although some ethical frameworks exist.
- *Data responsibility issues*—including protecting data from misuse and respecting refugees' data-related rights—are growing more urgent and complex as aid operations create and collect increasing amounts of personal data.
- *Bias* is introduced or exacerbated by technology-based humanitarian efforts when they exclude certain groups or perpetuate inequality or discrimination.
- Technology initiatives in refugee contexts might suffer from *conflicts of interest.* Some of the most common goals driving such initiatives include benefiting refugees, improving the operations of aid groups, and testing a new technology to meet organizational or personal objectives unrelated to the best interests of the refugees. In order to weigh risks, it is important to have clarity about motivations, interests, and intended results.

When humanitarian organizations fail to account for potential problems in these four areas, technology initiatives can subject refugees to security and privacy threats, discrimination, and risks related to technology experimentation. Ethical, security, and privacy challenges have direct consequences for both refugees' well-being and overall levels of trust in the humanitarian system.

Conclusions and Recommendations

Technology is changing the roles and responsibilities of refugees, aid organizations, and technology developers in responding to refugee crises. As these roles and responsibilities evolve, there should be better coordination in the strategic investment in and use of technology, which should lead to more opportunities for private-sector engagement and improved aid operations. Investment in technology in refugee settings is often made without preparing for the full life cycle of the technology, and the technology is sometimes implemented in advance of needed ethical and security frameworks. Furthermore, there has been sizable investment in creating specific apps for refugees, many of which fizzle out over time. To address these concerns, we offer the following recommendations for stakeholders involved in developing or using technology in refugee settings:

- *Focus private- and humanitarian-sector technology investments more strategically, weighing risks and benefits and considering the full technology life cycle.* We offer a suggested framework, a *business canvas model,* for developing investment strategies that are driven by needs; address human and institutional changes; account for political, legal, cultural, and geographical barriers; balance risks and costs with derived benefits; and account for the different stages of the technology life cycle.
- *Invest in sustained and mainstream platforms, data standards, and digital infrastructure.* Such investments by donors and technology companies should enable aid agencies to make better use of existing technology rather than creating fragmented, underutilized, or unmaintained apps.
- *Plan for technology scale and phaseout.* Some technical solutions will succeed, while others may not be relevant in the future. Aid agencies should set criteria for phasing out solutions that have less impact and should reallocate resources appropriately.
- *Invest in internet connectivity, not new apps, for refugees.* When refugees had access to the internet and other technology, they made good use of that technology, relying on mainstream platforms. However, refugees have inconsistent access.
- *Improve the strategic organization of the technology ecosystem through a wedding registry approach.* For a large range of stakeholders, opportunities to partner and contribute effectively may become more apparent through the lens of a systematic framework addressing the entire technology life cycle.

- *Improve technical capacity in the humanitarian community.* To accomplish this goal, technology companies could pay for or provide training, and aid agencies could hire and better train staff with technological skills. In addition, donors could provide startup or maintenance financing, and consortia could support training and platforms and donate time and skills.

- *Improve effectiveness and security in data management.* Data guidelines should be developed at the United Nations level. Regional data management plans based on these guidelines should be developed and implemented, and risk analyses should be conducted periodically, balancing the benefits of retaining vast data sets with the risks of securing the data.

- *Develop an ethical framework for technology in humanitarian settings.* As part of the ethical framework, develop guidelines for evaluating the balance between risks and benefits in using new technologies in refugee settings.

- *Develop legal frameworks governing technology, digital identity, and financial access in humanitarian settings in host countries.* Laws and policies have not caught up to uses of technology, leading to either unregulated or prohibited uses of technology in refugee settings. Gaps in the ability of refugees to present identification documentation or access common digital money-management tools impede important aspects of daily life.

- *Develop an improved evidence base for technology in refugee education.* Educational tools are one of the main ways that private-sector companies have aimed to contribute to refugee situations, yet the evidence base for such tools' effectiveness is thin.

Through this study, we have found that there is a solid foundation of technology use in humanitarian settings serving a wide variety of needs, and multiple actors create a wide variety of solutions. What is often lacking is the ability to effectively deploy and scale solutions and maintain them over the long run. Fragmented and uncoordinated efforts lead to inefficiencies and do not allow for solutions to be reused across different populations and problem spaces. Future research can shed some light on these and other topics and guide donors, aid agencies, private companies, and NGOs to collectively provide better services and more access with fewer resources. And although technology will not solve the refugee crisis or even address its underlying fundamental causes, it is improving the lives and livelihoods of refugees worldwide and can do so to a greater extent in the future.

Acknowledgments

We owe a debt of gratitude to many people who have made this report possible.

Many thanks particularly to Tom Kalil at Schmidt Futures for his guidance and support for our work.

We thank all of the displaced people who contributed their important perspectives, time, and experiences with us during focus groups conducted for this study. These people were interviewed in Bogotá and Cúcuta, Colombia; Athens, Greece; Amman, Jordan; Pittsburgh, Pennsylvania, United States; and the Maheba Refugee Camp in Solwezi, Zambia. To keep identities confidential, we do not name these individuals. We also thank the following organizations in these countries for facilitating or conducting these focus groups for us: the Centro Nacional de Consultoría in Bogotá, Colombia; the Market Research Organization in Amman, Jordan; Ipsos in Athens, Greece; the Bhutanese Community Association of Pittsburgh, Pennsylvania; Jewish Family and Community Services in Pittsburgh, Pennsylvania; and Ipsos in Lusaka, Zambia.

In addition, we appreciate the generous time and insights of the following people who were interviewed for this study. In particular, we thank the following government and multilateral officials: Ariana Berengaut, director of programs, partnerships, and strategic planning at the Penn Biden Center for Diplomacy and Global Engagement; Ambassador Tom Fletcher, visiting professor at New York University Abu Dhabi; Andrew Harper, director, Division of Programme Support and Management, United Nations High Commissioner for Refugees (UNHCR); Keith Hiatt, chief, information systems management section, United Nations International, Impartial and Independent Mechanism (Syria); Mohamad Karnib, information management officer, United Nations Children's Fund in Lebanon; Bernhard Kowatsch, head of the Innovation Accelerator, United Nations World Food Programme; Jean-Laurent Martin, information management officer, UNHCR in Ecuador; Aswad Muhammad, head of information management in Erbil, Iraq, International Organization for Migration; Anjalina Sen, foreign service officer and senior refugee coordinator, U.S. Department of State; Karl Steinacker, team leader for digital identity, UNHCR; and Moises Venancio, regional adviser for Iraq, Syria, Egypt, Jordan, and Lebanon at the United Nations Development Programme's Regional Bureau for Arab States.

We also thank the foundation, academic, nongovernmental organization, and private-sector representatives interviewed: Leslie Aizenman, director of refugee and immigrant services, Jewish Family and Community Services of Pittsburgh; Rosa Akbari, senior adviser, Technology for Development, Mercy Corps; Alexander Bertram, technical director, BeDataDriven; Giulio Coppi, digital specialist, Norwegian Refugee Council; Scarlet Cronin, senior director of private-sector partnerships, Tent Partnership for Refugees; Sasha Davis, deputy director of corporate partnerships, Mercy Corps; Jessica Fullerton, deputy director of quality and performance in the emergencies unit, International Rescue Committee; Josephine Goube, chief executive officer, Techfugees; Megan Hershiser, senior officer of institutional philanthropy and partnerships, International Rescue Committee; Ammar Kahf, executive director, Omran Center for Strategic Studies; Sasha Kapadia, director of markets and partnerships, government and development, Mastercard; Jane Meseck, senior director, Tech for Social Impact, Microsoft Philanthropies; Laura Nestler, global head of community, Duolingo; Brandie Nonnecke, founding director, CITRIS Policy Lab, University of California, Berkeley; Alice Obrecht, senior research fellow, Evidence, Innovation, Accountability, and Adaptive Learning, Active Learning Network for Accountability and Performance in Humanitarian Action; Prat Panda, senior manager, East Africa lead for Accenture Development Partnerships, Accenture; Meghann Rhynard-Geil, senior adviser, Technology for Development (Digital Communities), Mercy Corps; Aline Sara, co-founder and chief executive officer, NaTakallam; and Kim Scriven, fund manager, National Society Alliance, International Federation of the Red Cross and Red Crescent Societies.

Lastly, we would like to thank Charles Ries (vice president of RAND International, RAND Corporation) and Jane Meseck (senior director, Tech for Social Impact, Microsoft Philanthropies) for their reviews of the report, which helped us strengthen its presentation and analysis. We also thank Marjory Blumenthal and Maynard Holliday, senior researchers at RAND, for their facilitation and guidance of the research effort. We further thank Kristin Leuschner for her careful review of the communication style of the report and Laura Coley for her work on formatting the draft document.

Abbreviations

3D	three-dimensional
GPS	Global Positioning System
NGO	nongovernmental organization
SIM card	subscriber identification module
SMS	short message service
UN	United Nations
UNESCO	United Nations Educational, Scientific and Cultural Organization
UNHCR	United Nations High Commissioner for Refugees
UNICEF	United Nations Children's Fund
UNOCHA	United Nations Office for the Coordination of Humanitarian Affairs
USAID	U.S. Agency for International Development
WFP	World Food Programme

CHAPTER ONE

Introduction

In the past two decades, the global population of forcibly displaced people has more than doubled, from 34 million in 1997 to 71 million in 2018, according to the United Nations High Commissioner for Refugees (UNHCR) (2019b). The 2018 number is roughly equivalent to the population of France or the United Kingdom. These displacements have been driven by many factors, including wars and failing states with unsettled security situations. The statistics include refugees, internally displaced persons, and asylum seekers but not migrants displaced as a result of climate change or economic disparities. (See the box below for definitions of important terms used in this report.) All these factors, in combination with ongoing global conflicts, suggest that the coming years will see a persistent or likely increasing number of forcibly displaced people across the world.

A Note on Definitions

Technology: In this study, our primary focus is on *digital technologies* (often referred to as information and communication technologies) and systems that can be enhanced by digital technologies. Throughout the report, we refer to this combination in shorthand as *technology.*

Migrant: A *migrant* is someone who lives in a country in which he or she was not born. Refugees are almost always migrants, but most migrants are not refugees.

Refugee: There are multiple categories of people who have been forcibly displaced. These include the following (UNHCR, 2019b):

- A *refugee* is someone who lives outside of his or her country because of a well-founded fear of persecution and who has been granted special status and protection under international law. There are 25.9 million refugees worldwide.
- An *asylum seeker* is someone who applied for protection as a refugee but is still awaiting confirmation of status. There are 3.5 million asylum seekers worldwide.
- An *internally displaced person* is someone who has been forced to flee his or her home because of internal strife or natural disaster but has not crossed an international border. Internally displaced persons do not have the same protection as refugees do under international law. There are 41.3 million internally displaced persons worldwide.

Although this study focuses primarily on refugees, many of its findings are also relevant to asylum seekers and internally displaced persons. For simplicity, in this report, we use the term *refugee.*

The 1951 Refugee Convention, which was established to address the displacement of civilians after World War II, provides the international legal framework for how states manage refugees. Its overarching model is to provide humanitarian assistance until displaced people can return to their home states. But this approach has not met the needs of either the refugees or the states that host them. Most refugees (15.9 million globally[1]) live in a *protracted situation*, which UNHCR defines as "one in which 25,000 or more refugees from the same nationality have been in exile for five consecutive years or more in a given host country" (UNHCR, 2019b, p. 22). Once a situation has become protracted, refugees are away from home for an average of 26 years (UNHCR, 2016a). As the conflicts that created the refugee situation go unresolved, half of the world's refugees reside in camps, sometimes for generations, with inadequate opportunities for education, work, or simply normal lives. The other half of the world's refugees live in cities, which creates challenges for how their host communities manage rapid population influxes and often leaves refugees living in uncertainty for prolonged periods. In addition, 85 percent of all the world's displaced live in developing countries, which struggle to provide opportunities even for their own populations.

With growing recognition that these circumstances are not adequately accommodating the scope or scale of the problem today, the United Nations (UN) General Assembly in 2016 agreed on new core elements of a Comprehensive Refugee Response Framework (UNHCR, undated-d). And in December 2018, the UN ratified a new Global Compact on Refugees (UNHCR, 2018). What is new in these principles is an emphasis on the self-sufficiency of refugees (e.g., living in urban areas and holding jobs as opposed to staying in camps indefinitely while waiting to return home), providing assistance to host countries so they can include refugees in their public services, and helping refugees access resources and tools to return home.

The new principles and the overall momentum for change have created opportunities for new approaches to managing refugee situations, such as enhanced use of technology to solve problems in humanitarian settings. Refugees and the organizations that assist them have turned to technology as an important resource, and, at this critical juncture, technology can and should play an important contributing role. Technology has shown promise through such uses as supporting crisis management, facilitating the operations of aid organizations, and providing a means of communication among displaced people. Indeed, a UNHCR study found that, by helping refugees maintain social networks and access information and essential services, mobile phones and internet access are as critical to refugees' safety and security as are food, shelter, and water (UNHCR, 2016d).

---

[1]   In establishing the number of refugees in protracted situations, UNHCR considered only refugees under its mandate. This includes Palestinian refugees in Egypt but not those in Jordan, Lebanon, the State of Palestine, or Syria who are under the mandate of the United Nations Relief and Works Agency for Palestine Refugees in the Near East (UNHCR, 2019b).

Yet, although there are many uses of technology in refugee settings, there is a strong case to expand and improve those uses. With this study, we aim to analyze technology uses, needs, gaps, and opportunities for helping displaced people and responding agencies. And we explore options to more systematically develop and integrate technology in humanitarian settings, thus improving the use of such solutions to ease the management of this global crisis.

Study Approach

We relied on several sources of data and approaches for this report, as described in this section.

Literature Review

The literature review considered nearly 200 documents, including academic literature, reports by aid agencies, media sources, websites, and gray literature (i.e., unpublished or informally published working papers, white papers, government documents, and so forth). We catalogued technological innovations in refugee responses in the past decade; the main uses of technology by refugees; the way that UN agencies, governments, and implementing partners use technological tools; ethical frameworks for using technology in humanitarian settings; and the ways that technology is integrated into humanitarian operations. We considered such topics as personal communications among displaced people, data management by humanitarian agencies, dissemination of information to displaced people, educational and job-matching tools, personal identity and property rights management, distribution of humanitarian assistance, coordination among humanitarian agencies, displacement tracking, and smartphone and internet access and connectivity.

Semi-Structured Interviews with Stakeholders

We mapped roles of relevant stakeholders, decisionmakers, implementers, and innovators, such as UN humanitarian agencies, nongovernmental organizations (NGOs), refugee leadership groups, private-sector technology companies, and government officials. We selected a sample of interviewees with expertise, leadership, and responsibilities in humanitarian aid and technology based on a representation of various types of organizations, individual leadership in the field, and referrals from other interviewees. We invited these individuals to be interviewed via email or by introductions from others, when feasible; the interviewees did not receive incentives for participation. We conducted 30 semi-structured interviews using a standard protocol (question list), and we sought to gather a variety of perspectives but did not seek to treat topics exhaustively. Interviews covered organizational perspectives of main technology uses; needs, gaps, and opportunities in displacement settings; options to integrate technology into humanitarian governance structures; sustainable access to funding; and relevant technologies under development.

Focus Groups with Refugees and Internally Displaced Persons

We conducted focus groups with refugees, internally displaced persons, or both in Bogotá and Cúcuta, Colombia; Athens, Greece; Amman, Jordan; Pittsburgh, Pennsylvania, United States; and the Maheba Refugee Camp in Solwezi, Zambia. Table 1.1 shows more details about our focus group sample.

We conducted the focus groups in Pittsburgh (with facilitation from Pittsburgh's Jewish Family and Community Services and Bhutanese Community Association). With our oversight, partner organizations conducted two focus groups in each of the other settings, providing us with transcripts translated into English. These partner organizations were the Centro Nacional de Consultoría in Colombia, Ipsos in Greece, the Market Research Organization in Jordan, and Ipsos in Zambia.

All focus groups were conducted with six to eight people of mixed ages (from 18 to 75) and education levels (from illiterate to college educated). The main selection criterion was that participants used digital technology in some way, such as through a smartphone, a computer, social media, or the internet. Partner organizations used this criterion and convenience sampling to identify participants, who received $15–25 gift cards or cash as an incentive for participation.

Table 1.1
Distribution of the Focus Groups with Refugees and Internally Displaced Persons

Location	Focus Group #1	Focus Group #2
Bogotá, Colombia	Refugees from Venezuela, men and women	Refugees from Venezuela and internally displaced persons from Colombia, men and women
Cúcuta, Colombia	Refugees from Venezuela, men and women	Internally displaced persons from Colombia, men and women
Athens, Greece	Refugees from the broader Middle East (Afghanistan, Iraq, Palestinian territories, Syria), men and women	Refugees from sub-Saharan Africa (Cameroon, the Democratic Republic of the Congo, Guinea, Ivory Coast), men and women
Amman, Jordan	Refugees from Syria, men	Refugees from Syria, women
Pittsburgh, Pennsylvania, United States	Refugees from the Democratic Republic of the Congo, men and women	Refugees from Bhutan, men and women
Maheba Refugee Camp, Solwezi, Zambia	Refugees from the Democratic Republic of the Congo, men	Refugees from the Democratic Republic of the Congo, women

We designed the focus group protocol to solicit refugees' perspectives on their technology uses and needs; their ability to access technology in various settings; their data security and privacy; and their use of technology for communication, transit from their homes to new locations, education, employment, access to available assistance, money management, and identity management. Through the focus groups, we did not seek to collect comprehensive information but rather to gather a sample of information that would allow us to understand interviewees' general perceptions of technology in their current environments.

Analysis of Interviews and Focus Groups

We uploaded the interview notes and focus group transcriptions into Dedoose, a cloud-based software program that facilitates qualitative coding and data analysis. Our qualitative coding process followed established research procedures to ensure the reliability of the coding, including development of a codebook and a meeting to discuss and resolve ambiguities and discrepancies (Ryan and Bernard, 2003). Our codebook covered key research questions in the interview and focus group protocols, such as various uses of technology, business models for financing and developing technology, roles of different organizations, ethical and security considerations, and ideas for improving the current situation. This approach synthesized a range of perspectives from humanitarian actors and displaced persons.

Development of Findings and Recommendations

Next, we analyzed and synthesized results from the literature review, stakeholder interviews, and focus groups with displaced people. And, finally, through several internal team workshops discussing our findings, ideas that were suggested during the interviews, and ideas proposed in the literature review, we developed recommendations for multiple types of stakeholders.

Limitations

Readers should keep several limitations in mind when considering the key findings and implications of this report. First, the stakeholder interviews (in which we spoke to 30 people) do not necessarily represent the full spectrum of all stakeholders working with technology in displacement settings globally. There were doubtless important perspectives not captured here. Thus, study findings should not be interpreted as representative of all refugee circumstances globally. Second, interview data relied on the self-reports of stakeholders who participated voluntarily, and we have no independent means of verifying the accuracy of their responses. The interview data could also reflect respondents' own biases. Third, because they involved samples of displaced people based on circumstances in only six locations, the focus groups were not representative of all displaced people globally. At the same time, we think that, even if it is not an exhaustive analysis, our combined set of approaches here has enabled us to frame the issues and main challenges regarding technology in refugee settings.

Roadmap for This Report

The remainder of this report is organized as follows:

- In Chapter Two, we provide an overview of the roles and responsibilities of key entities involved in using or contributing to technology in refugee settings. The entities discussed are refugees, aid agencies, host countries, donors, technology companies, consortia, and universities and research organizations.
- In Chapter Three, we discuss the main uses of technology in refugee settings: internet connectivity and access, communication with family and friends, information for journeys, establishment in new locations, language, memory and record preservation, employment, education, management and coordination, distribution of assistance, data collection and analysis, registration, and identity management and digital identity provision.
- In Chapter Four, we discuss the findings from our focus groups with refugees, highlighting their perspectives on uses of and needs for technologies.
- In Chapter Five, we analyze the business models by which technology has been developed in refugee settings and lay out concepts for improved business models.
- In Chapter Six, we lay out ethical, privacy, and security issues with the use of technology in settings involving vulnerable people.
- In Chapter Seven, we offer recommendations for improving roles and responsibilities, streamlining business processes, and addressing ethical and security concerns.

CHAPTER TWO

Roles and Responsibilities of Entities Involved in Using or Contributing to Technology in Refugee Settings

Multiple entities are engaged in using or contributing to (e.g., developing, funding) technology in refugee settings, and the roles, responsibilities, and relationships among these entities are complex. In this chapter, we draw on our literature review and stakeholder interviews to describe these main entities and their roles and responsibilities (Table 2.1). In particular, we describe the roles and responsibilities of refugees, aid agencies (UN entities and their implementing partner NGOs), host countries, donors (e.g., donor governments and foundations), technology companies, consortia, and universities and research organizations. We also describe some related considerations, opportunities, and challenges, including how, in some cases, technology is changing some of these entities' roles.

Refugees

Refugees actively engage with technology in multiple ways and in multiple settings. In some ways, technology has enabled refugees to better help themselves. As one stakeholder interviewee noted, "I think where tech is really interesting is where it's empowering refugees to do more and make the most out of their situation." We identified three roles for refugees related to technology in refugee settings: using technology, communicating with aid organizations via technology to convey refugee needs, and developing technology.

Using technology. Refugees make extensive use of technology, particularly via mobile phone and common communication platforms, such as Facebook and WhatsApp. Of all refugee households, 71 percent have a mobile phone (UNHCR, 2016b). One interviewee explained,

> Refugees are doing fine with tech. . . . As long as they had a phone and some sort of data plan . . . and a way to charge the phone . . . . An Afghan boy showed me his Instagram and you could see his journey in reverse. You're in mountains, on a boat, in a desert. . . . So, they were really great at using tech in some interesting, innovative ways.

We discuss these uses in detail in Chapters Three and Four.

Communicating with aid organizations via technology to convey refugee needs. Several aid workers described how technology was evolving as the aid community and refugees engage with each other. One explained,

> One of the things that's been revolutionary for us with tech is that, in the past, the way we delivered info to refugees was a top-down model where . . . humanitarian organizations decided what they thought refugees need. . . . But through Facebook, we're able to create more of a two-way communication channel.

Developing technology. Refugees have been involved in limited ways in providing input for the design of technology intended to help them. NGO and UN officials in our interviews said that they consulted refugees about their technology needs through focus groups and smartphone-based surveys and hired refugees to work as data gatherers. Refugees have also developed smartphone apps that help other refugees navigate journeys, learn about services in new countries, and gain employment. Examples include Gherbtna (developed by a Syrian refugee in Turkey), Alfanus (developed by Syrian refugees in Turkey), Bureaucrazy App (developed by Syrian refugees in Germany), and Chatterbox (created by a UK-based refugee from Afghanistan) (Corbett, Frey, and Marjanovic, 2017; Pearcy, 2018; United Nations Educational Scientific and Cultural Organization [UNESCO], 2018).

Aid Agencies

UN organizations and their implementing partner NGOs are the key managers of refugee response situations, and they take on many roles related to technology in such settings. These include managing operations and assistance that draws on technology; collecting data, conducting analyses, and managing data security; setting terms and values regarding the use of technology in humanitarian settings; providing innovation in technology use to solve humanitarian problems; and scaling technologies.

There are roughly 100 specialized agencies, funds, programs, related organizations, commissions, and other entities in the UN system (United Nations Department of Global Communications, 2019). Although multiple UN entities are engaged in refugee settings, some key actors are the UNHCR, the World Food Programme (WFP), the United Nations Children's Fund (UNICEF), the International Organization for Migration, the United Nations Office for the Coordination of Humanitarian Affairs (UNOCHA), the United Nations Development Programme, the World Health Organization, and the International Labor Organization.

UN agencies, donor governments, and other funders hire NGOs to implement programs or projects. For example, UNHCR spends about 40 percent of its annual expenditure to hire partners (NGOs, governments, or other UN agencies) to implement programs or projects (UNHCR, undated-e). NGO roles in both camp and non-camp refugee settings include aid distribution, protection, logistics, shelter, health, water, sanitation, nutrition, and education projects. Some NGOs serve as resettlement agencies.

Managing operations and assistance that draws on technology. UN agencies and NGOs rely on technology systems to manage their operations both in the field and at headquarters. According to our interviewees and literature review, many of the systems used are typical of business offices, such as Microsoft Office, communication technology, videoconferencing, and text message distribution systems (Orange Business Services, 2018). There are also innovative tools specific to the aid community (discussed further in Chapter Three), such as ActivityInfo (an online tool to coordinate humanitarian activities) and registration systems that use biometrics. Because of these newly available tools, technology is changing aid agencies' capabilities, approaches to managing assistance, and mandates. Technology is also changing how food and housing aid is distributed, how information is used and communicated, and programmatic coordination among agencies. For example, the WFP has increasingly focused on distributing cash rather than goods, turning the agency into a banking organization in addition to a food distributor.

At the same time, interviewees described challenges related to the aid community's use of digital platforms to manage assistance. Multiple interviewees noted that these agencies have weak technical resources, including platforms and staff skills. One described "a bunch of uncoordinated humanitarian aid efforts that could have benefited from some basic, 20-years-ago tech. But they don't have it deployed and don't have budgets for it." Another noted, "At headquarters, you might have some pretty tech-savvy people, and in the field, you may have some rogue tech-savvy people. But there's a lot of in-between, where folks just don't know how to use the tech." Reasons given for weak technical capacity include low budgets, the fact that agencies operate individually and compete for funding, insufficient technology talent in the humanitarian sector, and a donor preference for funding direct programming rather than technology infrastructure systems.

Collecting data, conducting analyses, and managing data security. According to interviewees and the literature, aid agencies are relying more on data collection and analysis in their decisionmaking and program management, and their use of technology to analyze biometric data, data science, mapping, and social networks is growing. "We have to compare with 10 years ago, we now do more data analysis," said one aid worker. Another said, "the nature of coordination has gone from physical to online . . . and data has gone from being quite sparse and nondigital to being prolific." In Chapter Three, we describe the uses and approaches of data collection and analysis in refugee settings.

In many cases, data analysis has improved the ability to manage programs and make decisions, but there are also challenges. One interviewee noted the drawback that funder reliance on data reporting requirements for programs leads to uncoordinated repeated data collection:

> The result of that is massive and repeated data collection. Refugees will roll their eyes at you like you're the 12th person to ask me about this. . . . And they feel they have to talk to you. They are trying to get some help, and instead they get people asking questions over and over and over again.

Other challenges often discussed in relation the collection of so much data were proper data collection and data security procedures, which we discuss in more depth in Chapter Six.

Setting terms and values regarding the use of technology in humanitarian settings. The aid community sets principles for the use of technology in humanitarian settings. For example, in 2016, the UN declared the internet to be a human right (Howell and West, 2016), and UN agencies have undertaken several efforts to create additional principles. Chapter Six discusses this concept in more detail.

Providing innovation in technology use to solve humanitarian problems. Aid agencies fund research to address humanitarian problems through technology. Many UN agencies and NGOs have internal innovation units that aim to improve the use of digital technology and data analysis (United Nations Innovation Network, 2018). For the World Humanitarian Summit in 2016, one of the four themes selected was "transformation through innovation" (Scriven, 2016). Yet, in our interviews, a common criticism of the innovation ecosystem in the humanitarian sector involved fragmentation, lack of coordination, and duplication of effort. Although one interviewee noted that some duplication can be beneficial "because you don't know which [technology solution] is going to be successful," others said, "There are many small groups with small solutions, but they are not big enough, so they fizzle," and "this has sprouted a lot of fragmented and piecemeal efforts."

Scaling technologies. Successful technologies scale and spread across institutions and national contexts and from headquarters to field offices. Yet multiple interviewees mentioned barriers to scaling. First, the decentralized system of aid organizations enables flexibility but also reduces efficient scaling of solutions. According to our interviewees, procedures are unclear for moving a technological approach across agencies or from one country context to another. One interviewee explained, "While [agencies] share best practices, they also pursue decentralized localized approaches, because you can't move things easily among different contexts." UN and NGO officials noted that some of their systems are not connected or interoperable. Second, the short-term nature of refugee assistance projects has created limited incentives for agencies to build larger, system-level technological approaches at scale. One interviewee noted, "Although there are similar goals, there are not always those incentives to collaborate or coalesce around single solutions." Third, the headquarters and field offices have different incentives. One interviewee noted,

> At the headquarters level, everyone wants to be super techno-savvy. . . . On the field level, many field teams have seen these techs get trialed, see they have tons of bugs, don't work in areas with low connectivity . . . . There's the perception that some of these techs and approaches are just being pushed by senior management as a solution to . . . enhance their profile.

Host Countries

In addition to hosting refugees on their soil, host countries provide refugees with public services and set legal frameworks for the use of technology. As reported by UNHCR (2019b), 85 percent of refugees live in developing countries. The top ten countries that hosted the most refugees in 2018 were Turkey, Pakistan, Uganda, Sudan, Germany, Iran, Lebanon, Bangladesh, Ethiopia, and Jordan (UNHCR, 2019b). The United States had the largest refugee resettlement program in the world every year between 1980 and 2017.[1] In 2018, Canada took the most refugees for resettlement (Radford and Connor, 2019).

Providing public services, sometimes relying on digital technology. About 55 percent of refugees globally live in urban areas, not camps managed by the UN or other entities (UNHCR, 2019b). This means that host countries often provide refugees with public services, such as education, health care, sanitation, safety and security, and employment services. Host countries often take on great expense when increasing their public services to provide for these new populations.

Host governments rely on digital platforms to manage and operate public services in refugee settings. For example, Jordan's Ministry of Planning and International Cooperation and Jordan's Department of Statistics together mapped public services, hospitals, schools, and population distribution for refugees (Culbertson and Constant, 2015), and the European Union's European Asylum Dactyloscopy Database stores fingerprint data on asylum seekers (Schiemichen, 2018). But interviewees noted weak digital management systems in some host countries. To combat this, donor countries have provided funding to developing countries to improve the management of refugee issues, which in turn helps those countries improve digital systems in general. An interviewee described the situation: "Databases are not that effective. National digital infrastructure is not so great. So donors are looking at how to improve the national system writ large, to respond to refugees better."

---

[1] Resettled refugees, as opposed to asylum seekers, are referred by UNHCR or other NGOs and must have legal permission to enter the destination country.

Setting legal frameworks for the use of technology. Host country governments set legal frameworks and policies within their national boundaries, including those related to the use of technology and data. The countries' laws on data, security, and privacy direct who can collect and access data related to refugees. For example, the European Union has enacted the General Data Protection Regulation, which governs use of personal data and privacy. Turkey limits access to refugee data to the Turkish government and the UN. Interviewees described how differences in country contexts (laws, financial infrastructure, markets, government policy) lead to varying technology solutions, which affects the scalability of those solutions. As one interviewee stated, "I'm struck by the fact that there are different policies in different countries that have implications for how tech could be used." The literature has noted that, in many cases, legal statutes have not kept up with technology developments, leading to some applications of technology outside the scope of legal frameworks (Coppi and Fast, 2019). Another interviewee explained why it is not always possible to apply new registration, identity, aid distribution, or other technology platforms in multiple countries: "At the end of the day, you always have to negotiate locally with what you can do and what you cannot do."

Donors

Budgets for the UN system and NGOs come from donor governments and some large private or foundation funders. Because these entities provide funding, they also have influence in setting priorities for action, and they take leadership roles.

Providing funding. The top five donors to UNHCR in 2018 were the United States ($1.6 billion), the European Union ($480 million), Germany ($396 million), Sweden ($143 million), and Japan ($120 million) (UNHCR, 2019a). Donor governments also fund other UN agencies and NGOs directly, and foundations provide funding to UN agencies or NGOs in support of refugee responses. For example, in 2015, Chobani Yogurt founder Hamdi Ulukaya (with a net worth estimated at more than $2 billion in 2019; see "#1425 Hamdi Ulukaya," 2019) pledged to give much of his wealth to refugee causes (Associated Press and Boyle, 2015) and founded the Tent Foundation, which aims to convene the private sector to develop solutions in refugee settings. Google.org has provided $20 million to disaster and humanitarian responses since 2015, in addition to volunteer time from its staff (Google.org, undated). And Microsoft Philanthropies serves as a key partner in Net Hope, a consortium of more than 60 humanitarian organizations, which collaborates to improve the use of technology in humanitarian settings (Spelhaug, 2018).

Setting priorities. Donors' ability to create priorities through their funding choices, including technology investments, was mentioned in interviews multiple times. One interviewee noted that "coordination is actually forced by the donors." A repeated comment by multiple interviewees was the challenge of the donor cycle, in which many efforts are funded on an annual basis, which leads to short-term thinking in setting priorities for technology investment (discussed in more detail in Chapter Five). The interviewees described donors' reluctance to fund technology systems, instead showing more interest in funding pilots, startups, and apps, and the interviewees attributed this inclination to interest in "exciting new things" rather than investment in platforms. One interviewee described the

> need to educate the donors on what returns they can expect and how much they need to invest. Our experience is that it is easy to get funding for pilots for new stuff like [artificial intelligence] or blockchain. It is harder to get funding for long-term projects.

Another conveyed that donors are now expressing a preference to "invest in the ecosystem." Some interviewees reported that there was a growing awareness among donors about the need for longer-term investments. As one government official said, "Donors are getting over the app thing now. We are not loving new apps."

Providing leadership. Donors also have convening power and the power to lead with ideas. For example, at the request of some governments (such as the United States), the 2016 United Nations General Assembly held a forum on how the private sector could contribute to sustainable development and refugee situations (United Nations Global Compact, 2016). In 2017, the European Union hosted the Brussels Conference on Supporting the Future of Syria and the Region, which was attended by many Middle Eastern countries that host refugees; the session led to regional agreements on job creation and education for refugees. One government official said, "That was, in my mind, the role of government in an ideal world. Not only doing—helping them identify needs—but also then being the matchmaker . . . to build these pathways with the tech community."

Technology Companies

Technology companies have been playing a growing role in humanitarian and refugee responses. Aid agencies and refugees alike make use of the common technology platforms that these companies develop, and the companies create apps and platforms specialized to meet a particular need in refugee settings. Furthermore, technology companies offer expertise, funding, training, and other tools as part of their corporate social responsibility commitments. Several interviewees noted that the private sector was eager to help but that there were limited formal ways of incorporating their contributions. As one noted, "You have the competence and the spirit and the energy and the resources of the private sector who really want to do meaningful things, but they are being shut out of the process in lots of different ways."

Providing typical digital technology products and services. Technology companies provide typical services, such as Microsoft Office, social media platforms, internet and mobile connectivity, and operating services. From our literature review and interviews, we found that off-the-shelf products and platforms, rather than specialized apps, are most common in refugee settings.

Developing apps and platforms to meet a particular refugee need. The private sector has stepped up to develop initiatives specific to refugee situations, although there is a need to scale up further and expand partnerships (PwC, 2017; Tali, 2018). They have done so through existing contracting mechanisms, donated time and solutions, and hackathons organized to develop solutions (PwC, 2017). In particular, an estimated half of the private sector's engagement in the Syria response has been related to using technology for education (Tausan and Stannard, 2018). Chapter Three describes some of the specific technology solutions developed, and Chapter Five describes business models for developing these solutions.

Offering expertise, funding, training, and other tools. As part of their corporate social responsibility commitments, some technology companies offer skills, money, staff time, advisory services, and internet connectivity to assist aid organizations and others in humanitarian settings. One interviewee noted that technology companies "tend to be very active partners. . . . They ideally, especially the younger companies, like to see their technology utilized in addition to their funds and very often lead with their tech." One technology company interviewee said, "it's often an education process for [aid agencies] to understand that if we are going to work together, it's not just our money . . . but it's our expertise, our voice." For example, Google has hosted the annual Humanitarian ICT Forum and allowed staff to use company time for humanitarian work (Maganza, 2017). Mark Zuckerberg announced in 2015 that Facebook would "bring internet access to UN-coordinated refugee camps" (Gaffey, 2015). And Microsoft launched Tech for Social Impact, which aims to improve technology systems for nonprofits through discounted pricing or donated advising and services (Spelhaug, 2018). Technology companies also play an important role in helping aid organizations raise money—for example, through such crowdfunding sites as IndieGoGo and Amazon wish lists (PwC, 2017; Gaffey, 2015).

On the other hand, several interviewees described challenges with the incentives often present in technology companies' participation. One NGO said of technology companies, "They care about how much visibility they get by donating money to us, how they look good." Others described the pros and cons of donated time: Although technology volunteers could really push forward projects and contribute, their involvement could also create unsustainable ventures and require leaders to spend more effort managing the time of temporary workers.

Consortia

Like-minded stakeholders have formed multiple consortia around various topics related to technology. Such consortia include technology companies, other private-sector companies, universities, UN agencies, NGOs, foundations, private volunteers, and others. In addition to formal consortia, there are informal groupings of refugees, volunteers, and NGOs that interact on such platforms as Facebook and WhatsApp. Roles that consortia take in refugee settings are sharing information about technology use and needs, building digital capacity and coordinating technical resources, and generating technological solutions to problems.

One interviewee described an "outburst of information associations and NGOs working on so many different causes." On the other hand, one interviewee described consortia as sometimes falling into a typical trap: "You have to make sure that every single group is represented. It ends up being huge and too cumbersome to do much with." Of note are how many such consortia there are, including the following:

- The Tent Partnership for Refugees is a coalition of more than 100 companies aiming to help refugees through employment, education, and access to financial services (Tent Partnership for Refugees, undated).
- NetHope is a consortium of nearly 60 leading global NGOs and 60 technology companies and funding partners that seek to apply innovative approaches to development, humanitarian, and conservation challenges (NetHope, undated).
- The Humanitarian ICT Forum is an annual forum hosted by Google to address humanitarian problems with technology, such as digital payments, humanitarian data analysis, and communication (UNOCHA, 2017).
- The Smart Communities Coalition, co-chaired by Mastercard and the U.S. Agency for International Development (USAID), seeks to improve the delivery of essential services to refugees and host communities through public and private collaboration and technology (Mastercard, undated).
- Techfugees is an NGO that organizes conferences, workshops, and hackathons to generate technology solutions for displaced people (Techfugees, undated).
- ID2020, led by Accenture and Microsoft with other companies and nonprofit organizations, targets the UN goal of providing legal digital identity (Juskalian, 2018; ID2020 Alliance, undated).
- The Digital Humanitarian Network is a volunteer network that offers information and communication technology skills in emergencies (Betts and Bloom, 2014).
- The Rapid Education Action (or REACT) Initiative has more than 50 companies pledging time and ideas; communication services; educational content; free consultancy services on developing projects in Syria, Yemen, and Chad; logistics for education materials; and information technology support (Fletcher, 2017). The initiative, developed by the Global Business Education Coalition, aims to

channel corporate contributions in support of education in emergencies through partnerships among businesses, UN agencies, NGOs, national governments, and actors (Global Business Coalition for Education, undated).

- The Technology and Education in Crises Task Team, part of the Inter-Agency Network for Education in Emergencies, is a group of nonprofit, academic, philanthropic, and private-sector experts focused on enhancing educational opportunities during emergencies (Inter-Agency Network for Education in Emergencies, undated).
- The United Nations Innovation Network, co-chaired by WFP and UNICEF, aims to promote innovation, tools, and best practices in the UN system. The network comprises more than 65 entities (United Nations Innovation Network, undated).
- The Global Tech Panel (led by European Union High Representative Federica Mogherini) aims to bring together technology, civil society, and diplomacy leaders to address global challenges (European Union External Action, 2018).
- PeaceTech Lab aims to bring together the private sector, conflict management professionals, the technology sector, academia, and government toward reducing violent conflict using technology, media, and data (PeaceTech Lab, undated).
- Startup Boat coordinates entrepreneurs, scientists, NGO workers, activists, developers, artists, and investors to develop solutions for refugees in Europe (Mengel, 2018).

Sharing information about technology use and needs. According to our interviewees, these consortia share and publish information, provide updates, conduct webinars, coordinate, create communities around interest in technology in humanitarian settings, and promote building digital capacity in the humanitarian community. Some of the informal groups share information online about refugee-related laws, distribution of aid, employment opportunities, and how to navigate life in new countries.

Building digital capacity and coordinating technical resources. Some consortia come together to help build digital technical capacity and expertise among the humanitarian community—for example, through training or advising. As one interviewee stated, "So much of digital tech is about the people and processes, not the tech." Interviewees described how bringing multiple perspectives together with various types of expertise was particularly valuable in this context: "There is more and more need to have projects that are mixing UN, university, NGO, and private companies' abilities," and there is a need for a "bridge between refugees, NGOs, and technology, so they can work effectively together."

Generating technological solutions to problems. Consortia bring together members to develop innovative solutions to common problems in humanitarian settings. One interviewee described the target as "big problems we need to solve overlaid with where innovation can help us overcome it."

Universities and Research Organizations

Universities and research organizations play an important role in providing innovation and research on technological uses in refugee settings. There are many such initiatives, and we highlight just a few here:

- UNHCR and the Rochester Institute of Technology and International Relief and Development created RefuGIS, a project with refugees living in Zaatari Camp in Jordan, to learn and build the geographic information services needed in the community (Tomaszewski, 2018; Tomaszewski, Martin, and Hamad, 2017).
- Stanford University and UNHCR Innovation partnered to pilot ASCEND, a mass-message platform with refugees in urban Costa Rica (UNESCO, 2018).
- Stanford's Immigration Policy Lab developed an approach to match refugees to host cities, based on employment opportunities (Holder, 2018).
- The Center for Information Technology Research in the Interest of Society at the University of California in Berkeley studies ethics related to technology in displacement settings (Center for Information Technology Research in the Interest of Society and the Banatao Institute, undated).
- The Refugee Learning Accelerator project at the Massachusetts Institute of Technology supports computer scientists and engineers from the Middle East to create technologies for refugee education (MIT Media Lab, undated).
- One interviewee described an ongoing effort between UNHCR and Penn State University to develop qualitative data analysis approaches for text mining in refugee settings.

Conclusion

As outlined in this chapter, many entities are involved in using or contributing to technology in refugee settings:

- Refugees use technology, communicate with aid organizations via technology to convey refugee needs, and develop technology.
- UN agencies and implementing partner NGOs draw on technology in managing the overall refugee responses and programs for essential services, collect data and conduct analyses, set terms and values regarding the use of technology in humanitarian settings, provide innovation in technology use to solve humanitarian problems, and scale technologies.
- Host countries use technology in offering public services to refugees and setting legal frameworks governing the use of technology and data.
- Donors provide funding, set priorities, and provide leadership toward technological solutions.

- Technology companies provide products typical of other circumstances; develop apps and platforms specialized for refugee settings; and offer expertise, funding, training, and other tools.
- Consortia bring together different types of stakeholders to share information, build digital capacity and coordinate technical resources, and generate solutions.
- Universities provide research and innovation.

With all these entities that are engaged in some way with technology in response to the refugee crisis, changes in technology are resulting in changes to each entity's roles and responsibilities. At the same time, some of the people we interviewed perceived that more could be done to integrate technology in aid organizations' and other entities' operations, by addressing specific technology needs, improving business processes, and developing legal and ethical considerations. We address these issues over the next four chapters.

CHAPTER THREE

Uses of Technology in Refugee Settings

This chapter describes uses of technology among both displaced people and the humanitarian actors helping them. We focus our discussion on the primary uses of technology because a comprehensive description of all examples of use and application would be too vast to catalog here. Indeed, as UNESCO has pointed out, the large volume of such examples "has given rise to the development of meta-apps and meta-platforms with the sole purpose of providing an overview of the existing resources, such as Apps for Refugees and Refugee Projects" (UNESCO, 2018). An additional caveat to the discussion of the many uses described here is the prevalence of *digital litter*—that is, apps that were built to help refugees but were not sustained (Benton, 2019), leaving a trail of outdated information; broken links; and false impressions of rich, available digital tools. Digital litter could potentially even cause harm by keeping misleading information prevalent. Indeed, one study found that most of the 169 refugee technology projects launched in 2015 and 2016 were inactive by 2018 (Mason, 2018).

As part of this study, we grouped technology uses in refugee settings into the following categories: internet connectivity and access, communication with family and friends, information for journeys, establishment in new locations, language, memory and record preservation, employment, education, management and coordination, distribution of assistance, data collection and analysis, registration, and identity management and digital identity provision. In the following sections, we describe the benefits and challenges of these categories of uses for both refugees and aid agencies. Important issues of ethics, data security, and bias are examined in Chapter Six.

Internet Connectivity and Access

Refugees place a high value on internet and mobile connectivity: UN research has found that refugees dedicate up to one-third of their disposable income to remaining connected (UNHCR, 2016b). Internet use among refugees may help reduce postmigration stress (Mikal and Woodfield, 2015) and contribute to social inclusion (AbuJarour, Krasnova, and Hoffmeier, 2018). When asked why refugees prioritize smartphones and internet connectivity, one of our interviewees responded, "Emotionally, they have lost all of their capital. Their phone is all they have left. They reach out to get information, support." Another said, "To connect with their families, that's been number one. To connect with news of what's happening around them. . . . They live in unlivable shacks, but they have an iPhone."

Despite this value, the degree to which refugees and internally displaced persons have access to internet and mobile connectivity varies. UN studies have found that, although 71 percent of refugee households own at least a basic phone, just 39 percent have an internet-capable phone (UNHCR, 2016b).

In some cases, humanitarian providers directly provide mobile connectivity or hardware. Examples of initiatives from the literature and our interviews include providing Wi-Fi in camps, phones or SIM cards (subscriber identification modules) to refugees, a power supply for charging devices in camps or on transit routes, and books through e-readers (UNESCO, 2018). One interviewee explained that some refugee resettlement agencies in the United States give refugees smartphones for limited periods "for [the agencies] to be able to reach them, for their safety, for 911, so they can contact family here."

At the same time, there are barriers to refugees' ability to take advantage of internet connectivity. Such barriers include a lack of knowledge of how to use digital technology, language differences, insufficient money to pay for connectivity, absence of a stable power supply, and location (urban refugees generally have higher levels of connectivity and technology access than rural refugees do) (UNESCO, 2018). One of our interviewees stated, "A challenge has been [refugees'] literacy level in their own language. It's not that high."

Communication with Family and Friends

The literature and our interviews and focus groups all indicate that one of the most important uses of technology for refugees is interpersonal communication—particularly for maintaining contact with, and finding, friends and family (Hounsell and Owuor, 2018; Mengel, 2018; Orange Business Services, 2018). One interviewee stated, "Families are spread out. . . . They depend on their telephones to maintain their families."

As noted in Chapter Two, refugees more often use long-standing or mainstream communication technologies, such as the telephone or instant messaging apps, than apps that are more complex or that are targeted specifically to refugees (Rutkin, 2016; UNESCO, 2018; UNHCR, 2016b). Simple communication apps, such as WhatsApp, Viber, Skype, and Facebook Messenger, are most popular. Our interviewees emphasized that these allow for international messaging and calling via Wi-Fi without requiring the user to have a phone number.

In addition, there are technology-based efforts that aim to assist refugees in finding their families (Mengel, 2018; Weiss-Meyer, 2017). For example, Refunite is a civic technology project working on family reunification in sub-Saharan Africa and Jordan. Additionally, the Red Cross's Trace the Face project posts photos of refugees who are searching for their families and allows them to be contacted via the program.

Information for Journeys

Refugees use mobile technologies to send and access information relevant to their journey from home. We identified the following specific uses.

Transportation and situational awareness for locations along the route. Maps—including images circulated through WhatsApp, Google Maps, or other apps based on the Global Positioning System (GPS)—facilitate the navigation of journeys and provide situational awareness during travel. Such functions have also enabled refugees to communicate their locations to seek help when in distress (Sebti, 2016; UNESCO, 2018). According to our interviews, smartphones with GPS functions enable refugees to navigate even in the case of language barriers.

Safety and security. Examples of apps that aim to improve safety and security for refugees' journeys are the Alarmphone project (a hotline to assist refugees in danger in the Mediterranean Sea), InfoAid (an app that assisted refugees transiting the Balkan Route by providing weather, transport, and border-crossing information), and the International Organization for Migration's MigApp (which provides migration-related information, including migration risks, global incident alerts, requirements for visas, and location-sharing) (Corbett, Frey, and Marjanovic, 2017; UNESCO, 2018; United Nations Innovation Network, 2018). However, such perceived support from technologies may also embolden refugees to make dangerous journeys and take risks.

Interaction with smugglers. Refugees have used technologies to reduce their vulnerability to extortion from the human smugglers they pay to help them during the journey. For example, some Syrian refugees created a system whereby a third party released payment for a journey in stages, using a code released upon the refugee's arrival at certain locations (Khalaf, 2016). At the same time, technology can be used as a tool for criminals to take advantage of refugees. For example, smugglers have used social media to pose as travel agencies, and others have sold fake documents through Facebook (Khalaf, 2016). One interviewee explained, "Their phone . . . It has a risk to make them more vulnerable to smugglers, . . . but it also has great opportunities to them to get out of their constrained position."

Establishment in New Locations

Refugees have used mobile communication and social media to adapt in host countries, and humanitarian organizations have used these tools to provide assistance in host countries (Butcher, 2018; Corbett, Frey, and Marjanovic, 2017; UNESCO, 2018).

Facebook has served as a particularly important source for refugees to find a variety of post-arrival information regarding employment, housing, food, events, and more. Targeted apps and websites also aim to provide host country or community-specific information about legal rights, resources, and assistance. Examples include Refugee. info (available for several European countries); Alfanus, Gherbtna, and the Services Advisor app in Turkey; the Mojaher App for Afghans in Iran; Refugee Info Bus in France and Greece; the Bureaucrazy app, Integreat, and the Welcome to Dresden app in Germany; and Ref-Aid in various countries. Additionally, chatbots have aimed to enhance the communication of information and support to refugees; examples of chatbots include DoNotPay, which gives legal advice, and Karim, which provides Arabic-language emotional and psychological support (Tali, 2018). Yet a trade-off between speed and quality of information arises in such technologies. Apps with more-reliable or verified information may be slower and more expensive to update, while faster apps (e.g., WhatsApp and Telegram) lack verification and may not necessarily have accurate information (Raymond et al., 2016).

Refugees have also used various technology tools to find housing in their new locations. For example, Refugees Welcome International provides a housing matching service in various European countries, Canada, Australia, and Japan (Refugees Welcome International, undated; Gaffey, 2015). And Airbnb, partnering with various organizations, has used its platform to help house refugees temporarily via its Open Homes initiative (Airbnb Open Homes, undated).

Across almost all six of the host countries in which we interviewed displaced people, refugees reported little to no knowledge of existing technologies designed specifically for refugees, and when they did report knowledge, most said that the technology did not work or did not meet their needs. However, refugees did report that, when they heard about resources from other host countries, they desired these resources in their own situations. A Middle Eastern refugee in Greece stated, "it would be helpful to have an application like they have in Australia or in Germany like Refugee Welcome. They are trying to help refugees, for instance, to find a home, to communicate, to help them handle the situation." Another said,

> We would need this kind of application that gives information or directions, something you cannot always find easily. It's a new life and a new environment for us. . . . Even in Jordan, they have this application where they give you information in 3D or information about the language and a lot of other things.

Bhutanese refugees in the United States also perceived a lack of online resources targeted toward refugees and the acculturation process. When asked about applications designed for refugees, a focus group participant responded, "I don't think we have that." Another noted, "I heard there's one in Canada." Another noted hearing about an app through HIAS (a nonprofit aid organization) but thought that it was not much used because people did not know about it. One internally displaced person in Cúcuta,

Colombia, said, "there are places where people want information. . . . Applications for displaced groups do not exist. . . . There is no application that reaches you and gives you all the information, where to go, with whom to talk. That is missing."

Language

Language apps have proven important for refugees and humanitarian organizations alike. We identified the following uses.

Translation. Translation has been a particularly important use (UNESCO, 2018). Both refugees and aid providers use online translation services, such as Google Translate, and interpretation services linking refugees with live translators (for example, the iOS and Android app Tarjemly Live is available in Turkish, Arabic, and English, offering phone and text message translation) (Tarjemly Live, undated). Other services, such as TikkTalk, provide verified interpreters (TikkTalk, undated-a; TikkTalk, undated-b; TikkTalk, undated-c). One of our interviewees who works at a nonprofit used a service called Talking Points to text with refugees; the service was originally developed for teachers to communicate with multilingual families of their students. Finally, Translators Without Borders is a virtual network of translators ready to assist in humanitarian operations (Translators Without Borders, undated-a; Translators Without Borders, undated-b; Skirble, 2010).

However, these language services also have drawbacks. Although they can provide fast assistance, one of our interviewees noted that such apps could provide only basic help: "Translations, in some cases, don't make much sense and could be misinterpreted." There could also be concerns with translators' qualifications, ability to interpret context, and even trustworthiness. As one of our interviewees stated, "We don't know who these online phone interpreters are. . . . So all of that is suspect. And even when they are talking, are they even saying what we are saying?"

Language learning. Refugees also use various technological tools to assist with language learning. For example, refugees have used Duolingo, which provides free app-based language instruction (Duolingo, undated). After discovering that some of the most common languages learned on Duolingo were Swedish and German, the company built several courses for Arabic speakers in an effort to help refugees. As of 2019, English, Swedish, French, German, and Spanish were available for Arabic speakers, with other languages in development. Other efforts to enhance refugees' foreign-language learning include the Ankommen app, provided by the German Federal Office for Migration and Refugees, and the MoLeNET mobile learning program in the United Kingdom (UNESCO, 2018). Yet most language-learning apps support individual language practice and memorization and cannot fully replace in-person language courses, in which learners engage more fully with a language (UNESCO, 2018).

Language teaching. Refugees have also found jobs as language teachers using such platforms as NaTakallam and Chatterbox, through which they can earn income by serving as conversation partners via Skype for students who are learning their languages. Originally created to support Syrian refugees teaching Arabic, NaTakallam now offers language sessions in Arabic (with displaced persons from Egypt, Iraq, Palestine, Syria, and Yemen), French (with displaced persons from Burundi, the Democratic Republic of the Congo, and Guinea), Persian (with displaced persons from Afghanistan and Iran) and Spanish (with displaced persons from Venezuela and Central America).

Memory and Record Preservation

Refugees use technology to maintain memories and identity. Refugees use their smartphones to store and refer to photos and memories (Toor, 2017) and, according to our interviewees, to take photos of their passports and other vital records. As one of our interviewees stated, "Everything [refugees] have lost is in images and photos." Several studies found that refugees use mobile phones, Facebook, and other social media to help establish a *new* identity—one that preserves the individual's cultural past but also creates a sense of belonging in the new host society (AbuJarour, Krasnova, and Hoffmeier, 2018; Wilding, 2012). Refugees also document personal experiences and news in a form of citizen journalism, sharing their stories through digital and mobile storytelling (UNESCO, 2018; Maganza, 2017). For instance, the International Rescue Committee and YouTube partnered to share refugees' stories, UNHCR worked with Google on the Searching for Syria website, Voices Beyond Walls is a youth digital storytelling project for Palestinians, and the Children in Communication About Migration project involved refugee children in media production activities.

Employment

Online apps aim to help refugees find employment (Almohamed and Vyas, 2016; Bacishoga and Johnston, 2013; "Brisbane to Help Asylum Seekers and Refugees Resettle in Australia via Technology," 2017; Butcher, 2018; Transformify, undated). Examples include the Transformify Rebuild Lives Program (helping refugees connect with businesses), Refugee Talent (a platform in Australia), and JustArrived in Sweden (matching foreign-born job candidates with employers). However, one of our interviewees noted that such apps are usually too small to gain significant traction with employers; thus, the interviewee considered larger platforms, such as LinkedIn, to be more useful. Some host country governments also allow refugees to access national employment databases, such as İŞKUR (the Turkish Employment Agency) in Turkey.

Technology can also be used by refugees for self-education and reporting. One interviewee noted that instructional YouTube videos provide step-by-step videos on how to acquire a work permit and other useful processes in some countries. Refugees have also used technology to further entrepreneurship. One of our NGO interviewees pointed out that, in Turkey, female-headed refugee households are using WhatsApp to market their homemade goods. Another interviewee explained that refugees and others can report workplace issues or abuses online.

Education

Education-related technology is one of the most common uses we found in our literature review. With 60 percent of all refugee children out of school (they are five times more likely to be out of school than other children) (UNHCR, 2016c), there has been significant interest from humanitarian organizations, foundations, and technology companies in exploring uses of technology for education-related purposes.

However, although there are many examples of initiatives using technology in education, their effectiveness is mostly unproven (UNESCO, 2018; Vosloo, 2018; Wagner, 2017). The effectiveness of technology in any educational setting (not just refugee settings) is not fully understood. Furthermore, digital learning initiatives often do not lead to formally recognized educational certifications and therefore are viewed as less valuable to refugees. At the same time, because it is often hard to ensure the availability of classrooms and teachers in refugee settings, experiments with using technology for refugee education are appealing. Several studies highlight that, although a technology-based solution might support learning, such solutions work best in combination with an in-person component and learner-centered pedagogies, mentoring, community connection, and teacher support mechanisms (Tausan and Stannard, 2018; UNESCO, 2018). We identified the following uses for education technology in refugee settings.

Kindergarten through grade 12 education. In the ongoing Syrian refugee crisis, nearly two-thirds of Syria's population has been displaced, and only one-fourth of Syrian refugee children accessed kindergarten through grade 12 education in the early years of the crisis (Culbertson and Constant, 2015), although enrollment has since risen. As a result, the crisis has spawned multiple experiments with online curriculum-based education in primary or secondary schools (UNESCO, 2018). Educational initiatives have included efforts based on national curricula, such as a UN-developed effort and Elmedresa.org, both of which aim to teach Syrian children based on the Syrian curriculum in Arabic. Tabshoura offers materials based on the Lebanese curriculum through the online course management platform Moodle. And Nafham provides video lessons based on the Egyptian and Syrian national curricula.

Refugees have also used various offline learning programs, accessing content pre-loaded onto donated devices (such as books, videos from the Khan Academy, Wikipedia, and other educational resources) (UNICEF, 2014; UNESCO, 2018; Nuttall, 2014; Rumie Initiative, 2016; Thaki, undated; Wagner, 2017). Examples include the Raspberry Pi for Learning pilot project for children in Lebanon; the TIGER (These Inspiring Girls Enjoy Reading) program in Jordan's Zaatari camp, which provides girls with educational materials via tablets; the Instant Network School project in various countries in Africa, which provides tablets preloaded with resources; the Ideas Box program, which sets up digital educational resources in Burundi; Learn Syria and Thaki, which provide preloaded mobile or secondhand devices; and INGO Worldreader's provision of e-reader devices to Tanzanian refugee camps.

Supplemental educational content. Refugees access supplemental educational content via short message service (SMS) (i.e., a text messaging service available on most mobile phones), radio, and games (Dahya, 2016). Educational games aim to appeal to refugee children in such settings as the Middle East and Africa, often with the goal of improving literacy, math skills, and psychosocial well-being. Refugees in camps in Kenya have used Eneza's Shupavu291 service with lessons, quizzes, and teacher interaction via SMS (Otieno, 2017; UNESCO, 2018). And refugee children and others have received education via lectures provided over the radio (Carlson, 2013).

Postsecondary education. Refugees have very low rates of postsecondary education enrollment: Just 1 percent of refugees go to college, compared with about one-third of people globally (UNHCR, 2016c). Technological resources used by refugees pursuing higher education include massive open online courses offered through Coursera, edX, Udacity, Saylor Academy, and others (UNESCO, 2018). Some refugees have combined off-the-shelf versions of these open courses with on-campus programs, have received certificates from university partners of digital learning programs, or have completed online degree programs.

Technical and vocational education. Technology has both facilitated and been the subject of vocational training for refugees. Efforts—including in Kenya, Jordan, Egypt, Lebanon, Turkey, and Europe—have used on-site learning and online platforms to train refugees on such skills as data entry, coding, information technology, and hospitality (Agence Française de Développement, 2018; Butcher, 2018; Donahue, 2018; Khalaf, 2016; UNESCO, 2018). Examples of these platforms include Refugees on Rails, which teaches software skills; Power.Coders in Switzerland, which teaches computer programming and places refugees in information technology–related internships; and the ReBootKamp 16-week software-engineering course in Jordan.

Teacher training. Technology has facilitated teacher training, often supplementing face-to-face training initiatives. For example, in Kenya, the Teachers for Teachers project provides on-site training and mobile instant messaging to support teachers in the Kakuma refugee camp, and the Borderless Higher Education for Refugees project trains teachers in the Dadaab camps. Syrian teachers in the Domiz refugee camp in

Iraq have accessed digital materials through a cloud-based server as part of the Connect to Learn program. Massive open online courses have also provided teacher training (UNESCO, 2018).

Aid Agency Management and Coordination

Aid providers use various tools to coordinate and manage their activities, and multiple interviewees expressed that online tools are transforming the management and coordination of humanitarian operations. As one noted, "The nature of coordination has gone from . . . people in rooms to online." Another described the impetus for additional digital coordination tools: "So, you have a massive response and many partners working on the same populations. . . . Simply trying to map what everyone was doing compared to the needs was very challenging."

Like many other office environments, the aid community relies on typical business platforms, such as Microsoft Office, Tableau, geographic information systems, videoconferencing, and Skype, according to interviewees and the literature (Orange Business Services, 2018). In addition, there are tools that are specific to the aid community. A key example is ActivityInfo, a data collection and program organization software tool (ActivityInfo, undated). ActivityInfo serves as an information system to coordinate what various UN agencies and NGOs are doing in a particular emergency response. The increased transparency that comes with using ActivityInfo notwithstanding, one of the tool's weaknesses is that it relies heavily on whether stakeholders provide frequent and accurate inputs (Culbertson et al., 2016). In addition to ActivityInfo, individual aid organizations have developed their own software applications (Mengel, 2018). In the United States, some refugee resettlement agencies use Apricot Software, which was designed for case management and donor management for nonprofit organizations. Other efforts have sought to streamline the sharing of skills, lessons learned, and other information. For example, the Humanitarian Genome Project aimed to create a search engine for those at the operational level to quickly find information about evaluations and best practices in crisis contexts (Baker, 2014; Betts and Bloom, 2014).

Distribution of Assistance

Humanitarian organizations use technology tools in distributing assistance to refugees. We identified the following uses.

Communication about assistance. Aid agencies have communicated via SMS and radio to inform refugees about aid availability, timing, and logistics and to provide updates to those in camps about the supply of food vouchers, blankets, and heating fuel (Dette and Steets, 2016; Orange Business Services, 2018).

Cash assistance. In recent years, distribution of assistance has shifted away from providing physical items (e.g., food) toward providing cash assistance that refugees can use to purchase goods. Technology is supporting this change. The cash distribution systems rely on a selection of digital technologies that may include debit cards through banks, biometric identification (such as fingerprints or iris scans), and blockchain recordkeeping (Coppi and Fast, 2019).[1] The most prominent example is the WFP's cash distribution, carried out in collaboration with Mastercard and multiple banks. According to data that WFP provided directly to us, the organization distributed $1.8 billion of cash assistance in 2018 and may distribute up to $2.4 billion in cash assistance in 2019. WFP's Building Blocks project (piloted in Pakistan and also used in Jordan) relies on blockchain (Juskalian, 2018) and enables WFP to directly create virtual accounts on an Ethereum blockchain and upload funds to them. Aid recipients can use these funds to buy groceries or other goods, accessing their accounts via iris scans enabled by IrisGuard technology, which uses UNHCR biometric data to verify recipients' identities (Coppi and Fast, 2019; Kenna, 2017). The blockchain application records and confirms the transaction. In other examples, Catholic Relief Services and USAID have provided cards to refugees in Nigeria who had fled Boko Haram, UNHCR Lebanon provided cash through debit cards and vouchers, and the Finnish Immigration Service provided asylum seekers with prepaid Mastercards (Orcutt, 2017; Stulman, 2017; UNHCR, undated-a).

According to interviewees, cash is becoming the preferred model when the country context permits (e.g., when there are functioning food markets, and shops can accept debit cards). Relying on physical distribution of commodities is the option when markets are not functioning, such as in situations of a war, drought, or poorly functioning banking system. Cash provided can be either unrestricted (refugees have full discretion on spending) or restricted (e.g., to use only for food).

According to interviewees and the literature, benefits of using technology for cash assistance include stimulation of host community markets through refugees' spending, more-efficient use of funds because refugees can choose items (rather than selling unwanted items and using the cash to buy what they really want), decreased administrative costs, enhanced security against theft or fraud enabled by biometric verification, usability in multiple locations, and detailed transaction records (Balakrishnan, 2015; Coppi and Fast, 2019; Indrajit, 2017). One interviewee noted that using biometrics to support the distribution of financial assistance has "really simplified the whole process of distribution." Another described risk management:

> I think the donors have come a long way in quelling some of these fears because the reality and the assumption just haven't matched—the fear of it being funneled

---

[1] *Blockchain* is a public digital ledger of transactions that is cryptographically secured against falsification and unauthorized alteration.

to terrorists or spending it on vices. . . . Tech does bring some transparency to the mix. . . . People spend it on their basic needs, more or less.

Yet there are several challenges to these approaches, according to interviewees and the literature. For example, personal biometric data must be carefully managed and secured to avoid misuse. In addition, fingerprints as identification for card access may not be appropriate for everyone, such as some manual laborers who lack readable fingerprints. And there is significant complexity involved in the interoperability among systems. In the words of one interviewee,

> One product or partner provides, say, a biometric service. Another does straight data collection. Another just focuses on the payment and deals with the regulatory issues there. But all these techs need to work together and be connected and integrated. And so those investments to date are traditionally project by project, case by case.

Delivery of aid through three-dimensional (3D) printing and drones. Aid groups have employed such technologies as 3D printing (e.g., of prosthetic limbs) and drones to deliver supplies in hard-to-reach areas (Field Ready, undated; James and James, 2016; Refugee Open Ware, undated; UNICEF, 2017). Potential future uses of drones include extending and providing Wi-Fi or phone coverage (Weiss-Meyer, 2017). However, many applications of drones and 3D printing for physical provision of refugee aid remain in a pilot or aspirational stage. Challenges include concerns that drones distance aid recipients from providers and that people think of drones as military tools (Soesilo and Bergtora, 2016).

Data Collection and Analysis

Aid agencies use digital technologies to collect and analyze field and program data, registration data, survey responses, location information, qualitative data, texts, and other information. Interviewees noted an increasing reliance on data analysis to inform programming and decisionmaking. Yet interviewees also noted that these new capabilities have drawbacks, such as the overwhelming amount of data collected, ethical issues related to collecting so much data from vulnerable people (discussed in Chapter Six), and shifts in resources and time toward collecting and analyzing data. As one interviewee noted, "There hasn't been a broader harmonization or consolidation of the data collectors, so what you have is massive fragmentation. . . . There is a trend in funding to push for data-driven analyses of effectiveness. The result of that is massive and repeated data collection."

We identified technology tools being employed for the following types of data.

Program, survey, and location-based data. Some data tools are SMS-based, such as Ascend, piloted by UNHCR in Costa Rica, which sends mass messages or surveys to refugees via text (UNHCR, 2016b). Other tools are app-based. Some enable time and geostamping, as well as photo and video uploading (Dette and Steets, 2016). Other data tools track refugees' and displaced people's paths and circumstances. For example, the International Organization for Migration's Displacement Tracking Matrix publishes information about the trajectories and needs of displaced people (Displacement Tracking Matrix, undated). The World Bank and UNHCR established a joint data center that will collect demographic and socioeconomic information on displaced groups (World Bank, 2018). And UNOCHA's Centre for Humanitarian Data has developed the Humanitarian Data Exchange, which aims to consolidate and ease analysis of humanitarian data (Humanitarian Data Exchange, undated).

Educational data. Data systems on refugee education include information on schools, student and teacher attendance, student demographics, and student achievements. Examples of educational data systems include EduTrac in Uganda, which collects data via SMS; Kmobile Schools, which is a tablet and smartphone app designed to gather information on refugees attending school in Sierra Leone, South Sudan, and elsewhere; open-source systems, such as the UNESCO-developed OpenEMIS in Malaysia and Jordan; and Turkey's Education Information Management System for Foreigners, or YOBIS (UNESCO, 2018).

Data in hard-to-reach areas. Data tools also help organizations monitor programs in dangerous and hard-to-reach areas. Photo-sharing, screen-sharing, and Skype calls enable simulated field visits (Zikusooka et al., 2016). Drone and satellite imagery, as well as GPS-connected sensor and barcode data, facilitate situational awareness and remote monitoring of the locations of people and resources and the state of infrastructure, as well as the evaluation of program results over time (Dette and Steets, 2016; Raymond et al., 2016).

Big data. Aid agencies, the private sector, and others are also examining how public data and big data (including social media data) can assist in humanitarian contexts. For example, UNHCR has investigated how the analysis of Twitter feeds could provide insight into xenophobia directed at refugees (Orange Business Services, 2018). Facebook has used its data to help inform aid organizations about where certain types of aid may be required. The United Nations Office of Information and Communications Technology and the Internal Displacement Monitoring Centre worked together to put forth the #IDETECT challenge to crowdsource ways to use big data in displacement situations (Milanio, 2017; Ruhil, 2018; United Nations Department of Global Communications, 2017). And journalists and researchers worked on the Migrants' Files project from 2013 to 2016 to record deaths of migrants attempting to come to Europe (Fuchs, 2016; Migrants' Files, undated).

Qualitative data. A recurring theme in our interviews was the need for better tools to analyze qualitative data at scale. One interviewee expressed that the aid system is "struggling to deal with all this text" to analyze qualitative data in emails, text messages, program documents, and focus group notes. The purpose of analyzing the data would be to better understand lessons learned in an aggregated way and perceptions about security and other issues. One interviewee asked, "If you have 500 partners reporting monthly about what they are doing, could there be tools that would make that easier to process, summarize, turn into some kind of an information product for decisionmaking?"

Registration

The UN has used technological solutions for refugee registration. UNHCR's Profile Global Registration System (proGres) is an information technology registration and case management tool, rolled out in the early 2000s, that standardized refugee registration (UNHCR, undated-c; Goldstein-Rodriguez, 2004). In 2010, the UN began biometric registration linking fingerprint data with the proGres database (Lodinová, 2016). By 2015, UNHCR's Biometrics Identity Management System—which captures fingerprints, iris scans, and facial images—had completed development (UNHCR, undated-b). Additionally, UNHCR has an offline registration tool called Rapid Application (RApp) (UNHCR, undated-f).

Host countries also conduct biometric registration of refugees and migrants. For example, Turkey's Directorate General of Migration Management registers refugees using biometric data and manages electronic files in a database (Asylum Information Database, undated). In Europe, the European Asylum Dactyloscopy Database stores fingerprints of asylum seekers in a database that Europol manages and European Union member states can access (Schiemichen, 2018). Technology solutions have helped streamline backlogged registration and case management processes—for example, when portable iris scanners expedited registration procedures in Jordan, and Skype appointments expedited the registration of asylum applicants in Greece (UNHCR, 2016b).

According to interviewees, using technology to register refugees requires several considerations. First, registration data about vulnerable people fleeing conflict are sensitive and require careful protection; some governments restrict who can access such data. Second, in many cases, registration data systems do not connect easily with other systems, such as in case management or population tracking. At the same time, there could be security and privacy risks with linking multiple data systems that store sensitive personally identifiable information. Third, increased registration data can create a need for more staff to manage and analyze the data and coordinate with relevant governments, but low budgets do not always accommodate these needs. Finally, identities of people registered as refugees can be difficult to prove if those people do not have identity papers with them, which we discuss in the next section.

Identity Management and Digital Identity Provision

UNHCR explains the difference between registration and identity management: *Registration* is "the process of recording, verifying and updating information on persons of concern"(UNHCR, undated-c). *Identity management* includes registration but also involves identity validation or authentication of a claimed identity. However, many refugees flee their homes without documents that prove their identities, such as passports, birth certificates, marriage certificates, and educational certifications.

Technological tools can securely certify and store personal identification data. Nevertheless, operationalizing such tools in a scaled manner remains underdeveloped or hypothetical. Questions surround not only how digital identity *can* be provided but also how it *should* be provided and under what conditions. Multiple interviewees viewed digital identity as an increasingly important need but also uncharted territory, with a host of unanswered legal, ethical, functional, and security issues to resolve. As one interviewee said, "Those negotiations haven't taken place yet and they will be difficult. . . . This is a very multifaceted issue, and we are only at the beginning of it. We can't tell you where it is taking us."

Several interviewees pointed to proving one's identity as a form of human rights. Proof of identity allows refugees to access goods and services (e.g., register for education or health care), interact with the economy (e.g., get a job or obtain a bank account to be able to save or take loans), move freely (e.g., cross check points or obtain a driver's license), and sometimes even internet or mobile access (some countries require proof of identity to purchase a SIM card). Being locked out of financial systems is a particular challenge, as noted by the refugees in our focus groups. One interviewee described an emerging field of "forcibly displaced finance." The ability to verify identities is important to allow governments to manage security (for example, by knowing whether displaced persons have a criminal history) and to allow aid agencies to conduct case management and distribute limited assistance fairly, without duplication of resources.

However, there are risks inherent in creating multiple interoperable systems with data on refugees' identities. Data may not be properly protected, and, in some cases, being identified as a refugee can lead to discrimination. These issues are discussed in more detail in Chapter Seven.

Nevertheless, there are some digital identity efforts already underway. For example, UNHCR is developing the Population Registration and Identity Management EcoSystem (PRIMES), which aims to provide refugees with a legal digital identity that countries and businesses will accept (UNHCR, undated-f). PRIMES is how UNHCR pursues part of target 16.9 of the UN's Sustainable Development Goals to provide legal identity for all by 2030 (United Nations, undated). The UNHCR's Strategy on Digital Identity and Inclusion states, "Individuals will be able to request UNHCR to certify their identity"(UNHCR, undated-g). UNHCR intends for PRIMES to encompass and enable biographic and biometric registration and certification, case management, cash and in-kind assistance, and data management (including reporting and sharing) (UNHCR, undated-f). Thus, UNHCR will employ PRIMES for registration, distribution of services, and identity management and provision.

Beyond PRIMES, there are many ways that technology could be involved in managing and providing forms of digital identity for displaced people. For example, mobile authentication involves identity verification via a registered mobile device rather than through a physical identification document. Algorithmic analysis of a digital footprint provides options for inferring information about an individual, including identity, financial dependability, and more. A blockchain-enabled identification system could mean a back-end database similar to conventional systems, a transaction record connected to an identity already established elsewhere, or an "accretionary identification" amalgamated over time via transactions verified by other sources and stored on the blockchain (USAID, undated). ID2020, a public-private partnership involving such participants as Microsoft, also works toward target 16.9 of the Sustainable Development Goals (Juskalian, 2018). Existing systems—such as WFP's Building Blocks, which creates and holds digital records of financial transactions—could provide a type of credit history for refugees and others who may not otherwise have one, thus possibly enabling additional economic participation (Hempel, 2018; Schiemichen, 2018). Other efforts to create digitally portable documentation of educational credentials and skills include the Article 26 Backpack initiative from the University of California, Davis, and the Council of Europe's European Qualifications Passport for Refugees (de Leeuw and Skjerven, 2017); in addition, blockchain-based initiatives have similar goals to certify education and work experience (Butcher, 2018).

Despite these examples, current efforts to introduce digital identity management and provision in the humanitarian context are often small-scale, project-based, or aspirational (USAID, undated).

Conclusion

Technology can be used for a diverse set of purposes in the refugee context—for example, providing internet connectivity and access, supporting communication with family and friends, providing education and employment opportunities, facilitating distribution of housing and other resources, and providing a record of information about a displaced person's identity. At the same time, many of the apps and platforms developed for refugees have not been maintained over time, leading to digital litter on the internet with misleading or even potentially harmful information.

As described in this chapter, refugees place a high value on internet and mobile connectivity, which allows them to keep in touch with family and friends, maintain documentation regarding their identity and experiences, and access information and support. Technology-based efforts can assist refugees in finding their families, and some mobile technologies support refugees as they travel, helping them navigate routes and providing situational awareness. And refugees sometimes use technology to communicate with humanitarian organizations about refugee needs. Furthermore, refugees commonly use mobile apps to find housing, employment, and support services as they settle into their lives in a new country. Finally, refugees use technology to maintain memories and identity and often use social media to establish a new identity that preserves their cultural past but also creates a sense of belonging in the host society.

Aid agencies use various technologies to coordinate and manage their activities in supporting refugees. For example, ActivityInfo and other online tools allow organizations to coordinate their efforts. They also use technology to communicate to refugees about available assistance and to distribute assistance—which is especially important now that aid distribution has moved from a focus on physical items (e.g., food and blankets) toward an emphasis on cash assistance. The benefits of using technology for cash assistance include the stimulation of host community markets through refugees' spending, more-efficient use of funds because refugees can choose desired items to buy, decreased administrative costs, enhanced security against theft, usability in multiple locations, and detailed transaction records. Aid agencies also use digital technologies to collect and analyze field and program data, registration data, survey responses, location information, qualitative data, and other information from refugees, partner organizations, and publicly available sources. Finally, the UN and other aid organizations are using technology solutions to register refugees and, when possible, verify refugees' identity.

Refugees and aid agencies tend to use general and widely available technology (e.g., Microsoft Office, Facebook) more than technology developed specifically for refugees—although there are examples of both. As we discuss in Chapter Six, there are trade-offs associated with using common or widely available technology versus the technology targeted specifically to refugees. These trade-offs include convenience versus security, remote versus in-person connection, and accessibility and speed versus quality and accuracy. Key barriers to refugees' technology use are related to connectivity and access, literacy, language skills, and the efficacy of some of the technologies used.

CHAPTER FOUR

Refugees' Perspectives on Technology

In this chapter, we describe findings from our 12 focus groups in Colombia, Greece, Jordan, the United States, and Zambia. Although many of the issues described were similar across refugees in multiple contexts, some were particular to refugees who were living in a specific host country or were from the same country of origin.

In particular, we describe refugees' perspectives in three areas, which offer some personal views on many of the issues described in the previous chapter. We focus first on issues related to accessing technology, including the hardware and social media platforms used by refugees, as well as sources of internet connectivity and access. Next, we describe refugees' reports on their uses of technology for communication, information for journeys and establishment in new locations, language, education, employment, faith-based activity, health care, identity management, and money management. Finally, we discuss what refugees said were their concerns about technology, including issues related to (1) security and privacy and (2) reliance on digital technology.

Access to Technology

Hardware
Refugees across all host countries reported that they used mobile devices, primarily smartphones, more frequently than other types of devices, such as computers or tablets. The reasons given for this reliance on smartphones were their lower price compared with other devices and ease of use while on the move. A smaller number of refugees in Greece and the United States reported that they also used a laptop to access the internet; however, these refugees all reported that their phones were still their primary tool for accessing digital technology. Many refugees in other settings reported that they had previously owned a laptop in their country of origin or had a laptop with them that no longer worked.

Although most refugees believed that they could meet their needs by using a smartphone despite some of its limitations in function, others felt that the lack of access to a working computer was a limitation. One Syrian man in Amman said, "We would definitely buy a laptop if we were financially stable, since it is better for privacy." Several refugees across focus groups noted that some websites, documents, or video capabilities were not optimized to be viewed on a mobile device, which affected their employment opportunities or education. One sub-Saharan African woman in Greece described how not being able to access Microsoft Excel properly on her smartphone inhibited her freelance employment opportunities using her skills in accounting. A man in Zambia noted that it was difficult to continue his educational studies because he did not have a computer: "You can be trying to do your research on this phone, and you'll find that it's not successful. . . . I think it would be good for us to use computers for our studies because you can't study with a phone!" Refugees in Colombia and Greece described having some access to internet cafés with computers, which allowed them to print homework for their children, résumés, educational certificates and diplomas, and identity documents. Refugees in Colombia desired greater access to these cafés, and refugees in Zambia desired access to an internet café or computer center.

Bhutanese refugees in the United States described a federal program to provide phones to low-income families, which the refugees called "Obama phones." They reported that applying for the phones became more complicated as they began to earn more money. Furthermore, the phones had limited data plans, so the interviewees preferred to purchase their own smartphones when possible.

Social Media Platforms

In all six host countries where we conducted focus groups, the majority of refugees stated that Facebook, WhatsApp, or both were the most important and the most frequently used platforms, largely because of these apps' ubiquity, community groups for refugees, and low cost compared with voice and text message plans or landline telephones. YouTube and Google were the next most important and frequently mentioned, followed by Snapchat and Instagram. A sub-Saharan African woman in Greece stated, "When you want to connect with family or friends in Africa, they only know WhatsApp or Facebook." A Middle Eastern man in Greece said that he used Facebook "because some people I communicate with don't use any other app." One of the positive things about Facebook, as noted by a Venezuelan refugee in Bogotá, Colombia, is that "many people do not have a phone, but they have Facebook, because at any time they can connect through a phone or an internet café." One Syrian man in Jordan stated, "Facebook has become essential; you access Facebook on a daily basis."

As we describe in more detail later in this chapter, refugees used Facebook and WhatsApp to locate friends and family with whom communication had been lost; find and share information related to assistance, such as food, housing, and health care; advertise their skills and seek employment; find and use education services; enjoy entertainment, such as sports; find a marriage partner; shop for household goods, technology, and other items; and read news stories and current events. One man from sub-Saharan Africa described locating a friend after making the journey to Greece: "I was looking for him. . . . When I got my phone, I saw him online. So, I called him, and we eventually met." One Syrian man in Jordan described a Facebook page that helps Syrians crowdsource money for surgeries, and other Syrians in Jordan mentioned accessing UN information pages on Facebook. Syrian men and women in Jordan described "getting married through Facebook," and one Syrian woman in Jordan said, "you can even find a husband on the internet these days." Colombians and Venezuelans in Colombia mentioned promotions on Facebook for discounted or free food and subsidized rent for migrants in Colombia. Refugees in Greece mentioned information specific to refugees on Facebook, including about asylum processes and ways to travel to Europe. YouTube was described as a source for entertainment, cooking shows, and music. An illiterate Bhutanese woman in Pittsburgh described watching serial television shows from her home country on YouTube via her smartphone, after receiving lessons from her grandchildren.

Internet Connectivity and Access

As noted in the previous chapter, refugees greatly value their ability to access the internet. Yet the vast majority of refugees across all six host countries described limited or irregular access to Wi-Fi and cellular data. Common barriers were cost, lack of signal, lack of access to electricity to charge cellular devices, the number of people attempting to access a network, and lack of identification documents (which some countries require for purchasing a SIM card).

In particular, the cost of hardware, such as a router, to access Wi-Fi and the cost of data plans to access cellular data were barriers to internet access. Syrian refugees in Jordan noted that they calculated the cost of Wi-Fi into their rent because they considered it so essential to their daily lives. For refugees in Zambia, electricity outages were the most frequently cited disruption to service, closely followed by being too far from the mobile internet signal in the camp. One man in Zambia described how electricity problems raised the costs of his data plan (or "bundle"):

> It requires bundles and electricity to have internet connection. Because electricity here is a challenge, sometimes you may have all these things, and when it's time to connect yourself to the internet, the network starts breaking. And you end up by losing bundles just like that. It irritates me sometimes.

The inability to use devices because of connectivity problems was a source of frustration among focus group participants. A woman in Zambia stated, "some people do have tablets here, but it has become like decoration. They can't use them because there is no Wi-Fi here." One Middle Eastern man in Greece described the situation when he first arrived:

> In the beginning, when they put you in a reception camp and you have a lot of free time, . . . it's like you are in prison. At this point, you don't have internet. Technol-

ogy is not available. It happened to me. I had a lot of time, but I didn't have internet or a good phone to use it or do some work.

In several focus groups, refugees described difficulties accessing SIM cards for smartphones; they needed to change SIM cards each time they moved to a new country because roaming charges on their phones from home were high. And obtaining a new SIM card was sometimes impossible, given their identification status in the host country. A Middle Eastern woman in Greece commented, "In Turkey I didn't have any signal. I had to buy a SIM card and they wouldn't sell it to me, because I didn't have an ID card from Turkey. They have these rules, so I couldn't buy a SIM card. So, my phone was useless."

Some refugees share connectivity resources with each other. Most of the refugees living in Colombia, whether internally displaced or from Venezuela, reported that they shared access to Wi-Fi with family, friends, neighbors, or local business owners because the cost of owning a wireless router was too expensive for one person or one family. In the United States, Congolese refugees described sharing apps and software, such as Microsoft Office products. "Maybe I have one app and you don't have it, so instead of downloading it and paying, I send it to you easy. That's what Africa has been doing," said a Congolese man.

A small proportion of focus group participants reported that a lack of understanding about how to use the devices, applications, or websites and a lack of language skills were also barriers to accessing the internet. These refugees did not have experience with such tools in their home countries, and tools were in unfamiliar languages. A Bhutanese man in Pittsburgh said,

> There were no devices in Bhutan. A few educated ones saw television and computer in the camp. Many of us got the opportunity to learn technology, the use of computers and smartphones, once we arrived in the United States. Once we came here, we got to learn how to use the technology, and with that, we are able to contact our friends and relatives abroad, whether they are back in Bhutan, or in the camp, or those who have resettled in other countries.

A Congolese man in the United States described how language was a barrier to access:

> Another biggest thing I have here is the language barrier, because we don't know English. And here in this country, the phones are in English. And when you have to do something, you have to do it in English. If you want to listen to the news, the same thing. So anything you are going to do, it's in English. So the language is a problem to all of us I think.

Some of these refugees mentioned that they have enlisted the help of a younger family member or asked or paid someone to help them understand how to use their devices. An elderly Bhutanese woman in Pittsburgh said,

> I did not know the alphabet before coming to the United States. They started adult education here. So I have learned to use my iPhone, and what I have learned so far is out of asking the younger generation to teach me how to use those phones. And giving them a couple dollars in return.

A Congolese man in Pittsburgh said,

> Some of us, we have gotten to touch the smartphone [for the first time] when we are here in the United States. But we need to learn how to use them. . . . I need more lessons, more lectures, so that I can know how to use the phone. If I am to be included with others, I should know how to use the phone.

When asked broadly what could be changed to make life easier for refugees, several mentioned more access to digital technology hardware and better internet access.

Uses of Technology

Communication

The vast majority of refugees in all six host countries stated that they used digital technology most importantly and most frequently for communication with family, friends, and others. In more than half of the focus groups, refugees expressed a sense of duty to share some information—such as about access to services and assistance—online with a wider audience beyond family and friends.

Inconsistency of communication with friends and family while traveling to a new location was a source of stress for refugees and their loved ones. Almost all refugees reported limited access to digital technology, such as the ability to make calls, charge a phone battery, access GPS, or use cellular data, when traveling from their country of origin to their host country.

Information for Journeys and Establishment in New Locations

When internet connectivity was available, refugees made use of technology to assist with their journey from their home country and to settle into their new location. For example, the majority of focus group participants living in Greece and a smaller number living in Colombia reported that they were able to use Google Maps or other GPS services to navigate and cellular service to call for help during their journey.

Refugees in Greece and originally from sub-Saharan Africa or the Middle East were the only participants to explicitly comment on the role that technology played while hiring smugglers to take them to Europe. A sub-Saharan African refugee explained,

> I made it from Turkey to Greece. We used Google Maps for the locations. You know, once you're in the sea, it's very difficult to have signal. So what we do is that you go on Google Maps, you capture the location you are at, you send it back to somebody back in Turkey, and they let you know which direction you have to follow. So the person in Turkey is directing you once they see your location on the sea. Because it's not always easy for you to follow Google Maps when you are in the sea. . . . Because if you don't have that, you can easily get lost.

Another sub-Saharan African refugee said, "After two hours in the sea, I called 100. They said to me that I have to keep my phone on. They asked if I can speak English and I said 'yes.' . . . I kept the phone on and after 5 minutes, the big ship came."

The experience of sub-Saharan African refugees was mirrored in the process of refugees from the Middle East arriving in Greece:

> I came through the Aegean. I was on a rubber boat, and after 20 minutes, the engine stopped. Someone called the smuggler, and the smuggler called the police to come and save us. . . . Smugglers can be helpful. . . . Sometimes, if you don't have signal or Wi-Fi, the phone is useless. So it's important to always have signal when you travel. It helps you find the smuggler, but if you cannot find him, it's useless. . . . Also, you need to keep checking the weather.

Refugees from Venezuela talked about people helping and guiding them along the way, employing technology:

> In our case, obviously there is always a person who receives you. . . . You communicate by any means, by apps such as WhatsApp or Facebook, and that person . . . is instructing you, he gives you the address, he gives you everything.

Several Venezuelan refugees in Bogotá noted that they had access to digital technology while traveling: "Well, of all things, it worked well for me. All the way, it worked out for me," said one. These refugees primarily used the technology to communicate with friends and family and to navigate. However, they encountered issues with cell phone battery life and the ability to charge phones, given time limitations in travel or available electricity. They described needing to pay to charge a phone when the bus that they were on would stop along the route, but they were able to charge to only about one-third capacity because the bus did not stop long enough for a full charge. Several refugees described sharing signal or data with others along the journey.

In all countries holding focus groups, refugees described using transportation-related technology once they arrived in the new country for maps, bus schedules, and more. A Congolese refugee in the United States noted that technology helped his family navigate in their new location:

> My wife . . . got lost one day, and they just directed her using the phone. They say just do this, do this, do this, and you'll get home. And she made it home. And . . . when you want to wait for the bus, you can sit home and you know the bus is coming at this time, and it will not disappoint you. The same time, you go to the bus stop, you get your bus, and go away.

Many refugees lacked access to the internet during their journey because they lacked electricity, compatible cellular service, a data plan, a cellular signal along the route, or sufficient funds for a device or data plan. In addition, some refugees did not have access to digital technology on their journey or in their new location because it was not prevalent in their home country. Bhutanese refugees in Pittsburgh had not had access to technology when they fled a decade ago, and Congolese refugees in Pittsburgh described fleeing when someone in their village sounded a horn to communicate that they were under attack. Syrian women in Jordan stated that they all had mobile phones but were not able to communicate during their journey because they did not have a cellular signal. One woman said, "It was very difficult for us to come here, and my mother wanted to check on me when I was on the road, and she could not. I was out of touch for almost six days." A displaced Colombian in Bogotá related a traumatic journey, exacerbated by a lack of connectivity and electricity access:

> On the way, they can steal [your phone]. You can get assaulted . . . to get a 10-minute charge, to be able to load it. But maybe it does not work for you on that trip of at least two days. . . . Then it does not work because obviously while you are moving . . . the signal varies a lot, probably from bad to lousy.

Language

In ten of the 12 focus groups, participants mentioned using digital tools to learn English, even in countries where English was not the primary or official language. One Venezuelan refugee in Bogotá, Colombia, noted that the reason was employment: "For work, there are many places where English-speaking employees are needed, or in restaurants they are also needed." Other focus group participants mentioned learning Arabic, French, Greek, Hebrew, and Turkish. One Congolese woman in Zambia said that digital technology helped her learn English where shyness about being in a classroom might otherwise have prevented her: "I am a mother. I don't speak English well. I feel shy to put on a uniform and go to school to learn. I have an application on my phone that helps me to learn English." A Syrian man in Jordan stated that he used YouTube because "sometimes I search for things like teaching children pronunciation for Arabic words." Refugees also used translation services through Google Translate or other applications and through telephone translation services. Refugees in the United States noted that it would be helpful to have access to more real-time translation.

Education

Refugees across all focus groups reported using social media, smartphone apps, and websites run by private companies, nonprofits, and governments to take courses and learn skills. The focus group participants noted the importance of such resources specifically for those who were displaced or in remote areas. A Venezuelan in Cúcuta said that digital technology "makes it easier for people to study because suddenly they can be displaced and they have no means, they have no possibilities."

Use of Technology for Adult Education

Adult refugees stated that they used digital technology to learn skills and information and to obtain diplomas and certifications through online courses. Syrian women in Jordan mentioned that social media could be used as an educational tool: "Syrian girls usually get married when they are young, and this does not allow them to continue their education. And that is why using applications such as Facebook might help them in this regard." Another Syrian woman in Jordan said, "I believe that it also lowered the functional illiteracy rate, especially in elderly people since they are using their mobile phones to read and browse."

Courses taken through YouTube and other social media entities, as well as government-run services, were considered helpful for learning new skills and keeping up to date in one's professional field. One sub-Saharan African woman in Greece stated,

> I use the tablet for some personal research. For instance, we refugees don't have access to schooling. I studied medicine, so I keep taking courses through edX. It's online courses that you don't have to pay for. So you subscribe, and they send you the reading material. You can also take some exams to see whether you are moving forward, or you don't understand the subject.

When asked about plans to take courses online, Middle Eastern refugees in Greece described preparing for exams, taking a course in the history of civilization to both learn the content and practice English, studying language courses daily on YouTube, and learning medical skills.

Refugees were mixed in their opinions about whether it was better to take a course online or in person. Some liked the ability to take courses at their own pace online and in any location, while others noted that the lack of a teacher in the room hindered mastery of content. Middle Eastern refugees in Greece described how online courses could be taken to accommodate work schedules and liked the low cost and the ability to replay videos to improve comprehension. Other benefits mentioned by refugees in Zambia were access to an online library and the wide availability of materials.

Use of Technology for Primary and Secondary Education

Adult refugees across all host countries reported that their children use digital technology, the internet, and social media for educational purposes. They especially valued this access when their refugee status posed barriers to formal education. One Syrian man in Jordan commented, "I have an 11-year-old girl who did not go to school for a while when she left Syria, and she was able to learn everything she had missed from using YouTube." In addition, refugees described how online educational resources helped them adapt to their new host country. A man in Zambia noted,

> Internet helps me educate my children. Part of my children's education is from the internet because there are things like alphabetization in English. Through the internet, I teach my children and my wife so that we can be useful in this environment where most of the people speak English.

In Colombia, refugees from Venezuela and internally displaced Colombians reported that they often used internet cafés to access computers and print homework and other materials for their children. Refugees also reported using digital technology and hardware in creative ways. For example, a Syrian man in Jordan noted that access to a cell phone keyboard was helpful in teaching his daughter letters of the alphabet in another language.

However, despite these uses, some refugees still believed that it was best for their children to attend school in person. Syrian men in Jordan stated, "I would say that going to school is better," and "I think the traditional way is better." Refugees in Colombia and Zambia worried that technology was making children unfamiliar with books.

Employment

Refugees used digital technology to seek employment opportunities, follow job trends, keep up to date with skills important to their careers, and pursue self-employment and entrepreneurship.

Many refugees use social media sites to find job postings. Facebook was mentioned most often, but refugees also mentioned other country-specific sites, such as LinkedIn in the United States and OpenSouq in Jordan. In most of the focus groups, refugees mentioned specific online groups to help refugees gain employment. Two female Congolese refugees in Zambia described finding information through an internet posting: One woman in Zambia said, "I can be looking for a job, and one of my friends is working somewhere where there is a job opportunity. It's easy for that person to send me that application." Another woman in Zambia said,

> Employment adverts are on the internet. It helps when you are connected to the internet and you will be able to see it and apply online. But if you don't have access to digital technology, how are you going to know about the advert and how are you going to apply?

On the other hand, several refugees in Colombia found that it was easier to find employment through their interpersonal connections rather than social media or other online tools. One Venezuelan man in Bogotá was suspicious of online recruitment for jobs because he did not know how his personal information would be used by the recruitment site.

Several refugees referenced using the internet and social media to develop or maintain skills that might help them obtain a job. One sub-Saharan African woman in Greece stated, "In my country, the way we write our CVs [curriculum vitaes] is very different from how they do it in Greece. So, if you want to write your CV, you need to see the examples of the CVs here. . . . You can find templates in the internet." One Syrian man in Jordan noted, "I have friends who work in the same field as mine and they are from Morocco and Africa. They have channels on YouTube, and I usually follow them to view their designs. . . . I follow them on YouTube and sometimes I also post my work." One Middle Eastern man in Greece described how using the internet made him feel more comfortable about using his knowledge in his career field while observing safety practices:

> I used to work in the construction business back home, but things are different here. They use different materials. With internet, I learned how not to be afraid. . . . It helped me to start working here. I work now, painting houses. I learned how to use the plaster here. In my country, you can have your hands in the plaster all day and nothing happens. If I do it here, my hands are going to be burned.

Other refugees who needed a specific type of software to perform tasks for employers struggled to use their phones to accomplish the tasks without a computer. Several sub-Saharan African men in Greece described needing to use some complex smartphone apps. For example, an electrical engineer who previously used AutoCAD switched to Astrolabe, which he thought worked better on the phone, and an architect who previously used Archicad now uses other apps, such as Photoshop and a video converter. "It's great! I make business with this application," he said.

Faith-Based Activity

Several refugees discussed using digital technology for religious purposes. Female and male Syrian refugees in Jordan noted that digital technology increased their access to the Quran and that this technology made it easier to learn than using a physical copy did. A Syrian woman in Jordan said, "my brother used an application to help him memorize the Quran—voice application, since he did not know how to read and write." A Syrian man in Jordan said, "I would prefer my children to memorize [the Quran] using an application rather than going to the mosque or other religious classes as it is more reliable." In Zambia, a woman described similar access to the Bible: "Pastors don't carry Bibles; they have Bibles on their tablets." Venezuelan refugees in

Bogotá noted that they learned about opportunities to receive assistance (e.g., groceries and clothing) at local churches through social media.

Health Care

Almost all the information that refugees gave about health care and digital technology was related to methods of seeking care outside of traditional or official networks. Several refugees mentioned using Facebook or WhatsApp to find a doctor who could speak the same language or who agreed to treat refugees, especially given concerns about identification. A smaller number of refugees reported using technology to make an appointment with a doctor by calling or using an online portal. Syrian women in Jordan described learning through Facebook about a "free medical day for all Syrians." Syrians in Jordan also described crowdsourcing funds for surgeries and locating excess medications that were being given to those who could not afford them. And many refugees reported using the internet for self-care (e.g., self-diagnosing a health problem or finding remedies to that health problem) or for information about treatments that they could purchase at a pharmacy and avoid visiting a doctor.

Identity Management

Many refugees described using digital technology, particularly their smartphones and cloud-based resources, to save, share, and acquire documents related to their personal identities and educational or professional qualifications. A man from the Middle East noted that technology helped him while traveling to safety in Greece because, "before you make any movement, you can save your documents." Another Middle Eastern man in Greece stated, "For example, I studied art, but I don't have my certificate. I have photos, though, that can prove that." A Congolese man in Zambia echoed the same sentiment about saving his data through digital technology: "[Technology] helps to conserve my data. . . . It is easy to keep information on the Google Drive, which makes it easy to share it. Especially when it's saved on the phone." One Syrian man in Jordan commented,

> I was able to renew my Syrian driver's license without having to go there. . . . I met someone here in Amman and I gave him the old license and he renewed it back in Syria and then I paid him for it. I believe that you can find whatever you want on Facebook.

Refugees sometimes could be subject to scams when trying to use the internet to acquire identification. A Venezuelan refugee in Colombia reported that he was taken advantage of when attempting to apply for identification online:

> I sent a picture, personal information again, and sent $100. The next day, I should go to some place, to an address for the identification card. I went to the address

and it never existed. Because of the cell phone and technology, I was scammed on the internet.

One man from Guinea described how Greek authorities verified his nationality when he arrived:

> I was surprised when I arrived in Samos, . . . they ask you which country you come from. And they show the different kinds of money to you, and they ask you which one is yours to make sure that you actually are from the country you said that you are from. And then, they asked me where I lived in Guinea. I told them the area. They showed it to me on the screen, but I couldn't see my house on Google Maps. But there is a big pool close to my place and I saw that, so I showed them where my place is.

Money Management

Focus group participants' ability to use electronic money management tools was mixed; some had access, and some did not. Even where such services were available, refugees across almost all six host countries described money management as challenging. Refugees in most countries other than the United States noted that, without official identification that was accepted in their host country, they were not able to open bank accounts, apply for and use credit or debit cards, or use electronic payment services. These obstacles caused several problems.

For example, the lack of a credit or debit card or the ability to send payment online was sometimes a barrier to online shopping and online education courses, among other services. One Venezuelan refugee in Bogotá, Colombia, stated, "I got into the Facebook page of an English course and there I registered and sent the application, . . . [but] it was currently all online and I had to pay 67,000 pesos but with a credit card here in Bogotá." The lack of the credit card prevented the refugee from taking the course.

A smaller number of refugees stated that they had received a payment card that enabled them to make some limited purchases. A sub-Saharan African woman in Greece said, "They give us a refugee card we can use to buy things. We can use it at the supermarket, but you cannot buy things everywhere, and you cannot open an account. . . . It would be more convenient to use technology to manage money." Refugees living in Bogotá, Colombia, said that they carried cash to pay for goods and services "because we do not have access to anything [else]," as one explained.

Multiple refugees cited inconsistent services with other traditional means of transferring funds to family and friends, such as Western Union or MoneyGram. Venezuelan refugees in Bogotá, Colombia, noted that mistaken personal information could prohibit transferring funds to family in other countries. A Middle Eastern woman in Greece viewed Western Union as expensive but necessary.

The inability to use these financial services was a detriment to self-owned businesses or self-employment. To avoid electronic payment issues and fees, one Middle

Eastern man in Greece who was working as an artist found it easier to rely on friends physically traveling to other countries to deliver his artwork and return with cash; he struggled selling work in Greece when customers did not carry cash. "It would be helpful if there was another way for them to pay me," he said.

In contrast, refugees living in Zambia often had access to electronic money management and said that services were easy to use and enabled them to save money. As explained by one woman in Zambia,

> It's easy to send and receive money. You may be in need of money. . . . But you can be helped easily by someone who has got money and is far from you. That person can just send you money through mobile money and you receive within a few seconds. You can even pay bills online.

These same refugees noted that there were limits on their mobile money accounts and that they would need a bank account to use such services as Western Union or MoneyGram. However, they were unable to open a bank account in Zambia because of policies requiring identification. In addition, refugees noted that unreliable internet connectivity at the camp where they lived meant that sometimes they could not access electronic money management services.

Refugees in the United States reported the fewest problems in using digital technology for money management. Bhutanese female refugees in the United States reported that using social media and apps for money management was common: "I feel like all younger people like us use apps like Cash App or like Venmo," said one woman. Another described sending money via Facebook. A Bhutanese man explained, "I send money out to Nepal, Bhutan, wherever I have to send, to friends or—. And then through my phone, I send them the number; this is the number you need to take the money out."

Concerns About Technology

Security and Privacy

Many of the online security and privacy concerns described by refugees were similar to those of the general public (such as scams and improper use of personal data), although some issues arose specifically because they were refugees. Across all six host countries, refugees reported concerns about data security and privacy, as well as the need to protect themselves from fraud. In particular, they cited concerns about the privacy and security of the information that they shared online to prove their identity when applying for employment, education, or assistance; when completing online profiles; when managing finances or paying for services online; and when sharing other information. Refugees felt more vulnerable to fraud or data security breaches because of their refugee status and their unfamiliarity with the culture or language in their host country.

Refugees shared several examples of online scams or harmful use of personal data. For example, Syrian women in Jordan described coercion and hacking: "Usually we get links on WhatsApp that require us to share specific information, like our UN number, for example. And that is what we fear because the number is very important." Another in the group explained, "They can access your phone and photos and threaten you with some images in order to get the UN number or even to get money." Congolese refugees in the United States mentioned receiving calls from people who were attempting to steal their information or money. Refugees in Colombia and Jordan noted that they were concerned about their online security when responding to advertisements for employment and educational opportunities. Internally displaced persons in Cúcuta, Colombia, expressed that doing so could be "very ugly, very difficult because most are scams." Venezuelans in Colombia feared sharing personal information "because they can use it for scams" and "extortions." Other Venezuelans described receiving calls from strangers asking to speak to them, as well as their children, by name and asking for information. One internally displaced Colombian in Cúcuta explained,

> The truth is that you are very vulnerable with your cell phone, because you have all your data there, so many things, websites. . . . There are people who don't know and accept weird pages there; they install them on their cell phones and people extract all their information from them.

Refugee women mentioned vulnerability to harassment or similar problems online, such as pop-ups with sexually explicit information or photos. One woman in Zambia said, "There are many bad things on the internet, like pornography material. . . . It loses our dignity." Congolese female refugees in Zambia expressed a similar concern. One noted,

> It is exposing us. WhatsApp is a bit confidential, but Facebook is public. If you are on Facebook, it's very easy for someone to find your account. Someone needs just your name for him to search for your account. Especially if there is your picture on the profile.

One woman in Jordan stated, "Sometimes people use Facebook for the wrong reasons, and they might harass the profile of a girl and start messaging her in private—and this is annoying."

Young Bhutanese refugees in the United States noted that they worried about the need for better awareness of security concerns among people in their community, particularly older adults who had not used such technology in their home country. One stated, "I feel like the older generation in our community is not aware. . . . They're not really worried, or they don't really know about their security and privacy when it comes to technology."

Refugees described several strategies for dealing with these concerns. One strategy mentioned by Venezuelan refugees in Bogotá was to create partially or completely fake profiles on Facebook: "At least on Facebook, there is not a single fact that is true about me, nothing, nothing at all." They described posting their location on Facebook as Miami or Dubai. One said,

> Sometimes I prefer not to share my name or my last name, my identity on Google. If I am trying to make a payment and they ask for identity, I don't do it. I don't trust. . . . We know that sometimes they share our information with others, and I don't like that. I prefer to keep this information safe.

Reliance on Digital Technology

In general, refugees expressed mixed feelings about the presence of technology in their lives. As one Congolese man in Pittsburgh summed it up, "I am thankful for everything I get through the phone." One male sub-Saharan African refugee in Greece expressed a balance of views:

> It's like we say, one man's food is another man's poison. We can say the same thing about technology. . . . Technology can make you very intelligent. . . . [But] there will be some other people who can use technology for bad purposes. And we must be strict with our kids.

On the other hand, both male and female Congolese refugees in Zambia and Colombia expressed concerns about the negative side of relying on technology and called it "an addiction." One male focus group participant stated, "Some men spend more time on their phones than with their wives." A woman in Zambia commented that "it [eats] people's money; you may use the money that is meant for food and spend it on data bundles just [to] be able to download a movie." Similarly, refugees in Bogotá, Colombia, described the anxiety that occurs when they accidentally leave a phone at home and miss an important call from a relative, school, or work. Others complained about the "habit," and one noted "that bad vice that all these companies— . . . almost every human being gets addicted to that. Obviously, it does not give you time to have a break."

Conclusion

Refugees' perspectives on technology, as reported in our focus groups, offered personal insights into their uses of technology.

Access to technology. Refugees across all six host countries reported that they preferred smartphones over computers or tablets, and the majority of refugees stated that

Facebook, WhatsApp, or both were the most important and most frequently used platforms. In addition, refugees greatly valued their ability to access the internet, although the vast majority of refugees described limited or irregular access to Wi-Fi and cellular data. Barriers to using technology included the cost of data plans and hardware, problems accessing SIM cards, a lack of understanding about how to use the technology, and a lack of language skills.

Uses of technology. Refugees in our focus groups reported many uses of technology, which we grouped into the following categories: communication, information for journeys and establishment in new locations, language, education, employment, faith-based activity, health care, identity management, and money management. The vast majority of refugees in all host countries stated that they used digital technology most frequently and most importantly for communication with family, friends, and others.

Concerns about technology. Refugees reported being concerned about online scams and fraud, which mirror the concerns of most of the general public. However, refugees felt more vulnerable to fraud or data security breaches because of their refugee status and because they were unfamiliar with the culture or language in their host country. In addition, refugees expressed mixed feelings about the presence of technology in their lives, including the costs of technology and its potential to turn into an addiction.

Business Models for Developing and Deploying Technology in Refugee Settings

In the previous three chapters, we discussed the entities involved in providing technology in refugee contexts and the uses of technologies in those contexts. In this chapter, we shift the focus to examine the business models through which technology is developed and implemented in support of refugees and consider how these business models might be better applied when deploying such technology. We begin by noting that these models look quite different from the traditional business models used by private industry to develop and deploy technology solutions in support of everyday needs. The driving force in the private sector is profit, and, on the open market, customers ultimately decide what works and what is relevant.

In contrast, when it comes to deploying technology in support of refugees (or any other type of humanitarian assistance, for that matter), recipients of technologies have less of a voice, and technology development is heavily influenced by UN agencies, governments, and NGOs, who participate in a complex web of partnerships and relationships. In this case, the market is not the only or even the best adjudicator of success. Nonetheless, certain elements of the traditional commercial business model can apply in this context: identifying an opportunity; securing funding to develop the concept; demonstrating the concept; and deploying, scaling, and sustaining it.

We first explore what constitutes a business model in the context of deploying technology in support of refugees. We discuss the most common elements of a business model in this context and, drawing on results from the interviews, describe what seems to work and what does not. Next, we examine some barriers to and facilitators of developing and deploying technology innovations in refugee settings. Finally, we provide some constructs to think about how best to deploy technology in ways that meet refugees' and aid agencies' most pressing needs while also allowing private industry participants to understand the role they might play and ways they might even draw some financial benefit or other advantage.

What Is a Business Model?

There is no one widely accepted definition of a *business model* in the research literature. It is generally noted that business models emphasize a holistic, system-level approach to explain how firms conduct business and pursue opportunities (Zott, Amit, and Massa, 2010). However, different models take different perspectives on *how* a business creates value or *why* it creates value (Ritter and Lettl, 2018). To arrive at our suggested definition, we drew on some general concepts that are pervasive in the literature (Johnson, Christensen, and Kagermann, 2008) (see Table 5.1). For our purposes, we propose the following definition: *A business model is a set of four interlocking components (value proposition, key resources, key processes, and profit formula) that, taken together, create and deliver value.*

These four components provide a framework to understand the business models involved when developing and deploying technology in refugee settings. However, because private industry, nonprofits, and government agencies operate in fundamentally different ways, business models apply differently to each type of organization. In the next section, we discuss these components and what we learned from the interviews about how they are applied in the refugee context. We also include observations from our literature review.

Applying the Business Model Components to Developing and Deploying Technology in Refugee Settings

Value Proposition

The value proposition starts with the service or the need to be met, for either refugees or aid agencies. Across our interviews with stakeholders, we heard that value is derived not just from the technology itself but also from the way the service is provided and the end user's experience. Value can come from a technology's ability to meet a critical need and lower costs, as well as from such features as flexibility, familiarity, account-

Table 5.1
Components of a Business Model

Component	Brief Definition
Value proposition	The service or need one is trying to fulfill
Key resources	Anything that can be used to create value, including both monetary and nonmonetary resources
Key processes	Specific activities that need to take place for a product to be delivered to its intended users
Profit formula	Typically thought of as revenues minus costs—or more generally, as in the case of governments and NGOs, as cost-revenue structure

ability, and privacy. We heard about several initiatives that were focused too much on making the technology work and too little on providing optimal value *in the context of the need*. A value proposition should show how a service can be provided or a need met in a way that is better than what is currently available. For example, although the ability to lower costs was often mentioned, flexible pricing and modular services suited to the context also resonated with aid agencies. One interviewee described a successful subscription model that can be tailored to different price points, depending on whether it is used by a handful of users and a couple of administrators or several thousand users and many administrators. Other interviewees reported that technologies were more successful when their services were adaptable to a specific context and problem. Interviewees also emphasized the value of initiatives using existing platforms and familiar user interfaces. Better transparency and accountability were stressed, for example, in the use of blockchain and biometrics in Jordan. In other countries, privacy and data security were the key attributes providing value for certain refugee populations who feared persecution. For instance, we heard in an interview that "refugees from Myanmar . . . are extremely reluctant to provide [biometric data] because, in their experience in the country where they have run from, they think the data is used against them." We heard similar concerns regarding Syrian refugees.

Key Resources

Resources are anything that can be used to create value, whether monetary or non-monetary. In the context of developing and deploying technologies in refugee settings, a key resource mentioned in multiple interviews is a strong network of relationships, especially long-term relationships that can be leveraged to approach a problem over time and ensure that ideas and concepts are not only demonstrated but also deployed, maintained, and sustained. A strong network might include donors, UN agencies in host countries, refugees and refugee organizations, local governments, and other NGOs. Few, if any, organizations can do everything and do it well, so relationships are often critical to delivering value to refugees and aid agencies and achieving scale and long-term viability of solutions. For monetary resources, donors and grants remain a major source of funding for technology projects. On a much smaller scale, crowdfunding through such sites as IndieGoGo and Amazon wish lists offers another source of funding, and hashtags have helped with campaigns to raise funds (Gaffey, 2015).

Another resource mentioned in our interviews was an organization's image and level of trust with the people it is trying to serve. Such trust encourages people to stick with a solution, which is important when multiple organizations are deploying overlapping and often competing solutions. Interviewees also noted the importance of intellectual property and expertise. For example, we heard about Airbnb and other companies using existing technologies to serve refugee populations without having to reinvent a solution from scratch. Another important resource is knowledge and understanding of the local context, community, and refugee population. In addition, refugees, volunteers (e.g., employees from private companies who donate their time and expertise), and other people are invaluable resources. Finally, valuable resources that are often lacking in the refugee context are internet connectivity and "power at the last mile," which are often important in successfully deploying technology (see Chapter Three).[1]

Key Processes

Figure 5.1 depicts the activities and processes involved when a technology is developed, deployed, and eventually retired. In this section, we describe each of the six steps and provide examples as heard in the interviews.

1. Project Initiation and Concept Development

Each technology project starts with an idea, which can derive from either a market pull or a technology push (Souder, 1989). One interviewee described the distinction in this way: "Was it that people had a problem and they were looking for an answer, or was it opportunity-driven, like people see an opportunity outside the humanitarian sector, like drones, and say how could we apply that for humanitarian purposes?" We heard examples of both approaches.

The market pull occurs in several ways: Organizations conduct focus groups or surveys with refugees to understand needs; NGOs and UN agencies identify pressing needs in the field; organizational leaders at headquarters extract needs from field reports; and workshops or working groups come together to identify and prioritize needs. Interviewees also described observing the technologies that refugees use. As one interviewee stated, "Give refugees access to power, connectivity, and computers, and observe what they use them for and how."

Technology push can also occur in several ways. For example, pushes can be initiated by UN headquarters performing technology scans, by technology incubators identifying opportunities, or by NGOs identifying new technologies to address a

Figure 5.1
Key Processes for Developing and Deploying a Technology Solution in Refugee Settings

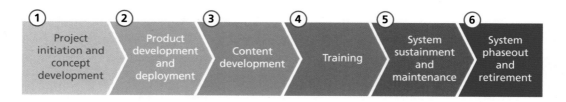

---

[1] The telecommunications industry coined the term *last mile* to refer to the final stretch of cables and other connections to deliver services to individual end users. As one energy executive explains, "People use the term to allude to the cost and effort of connecting individual homes and businesses to the telecom network, which was literally done by laying the 'last mile' of wire that went from the street poles (the network) to the home" (Kennedy, 2017).

specific refugee problem. Technology pushes seem to be favored by both NGOs and donors. Donors tend to be attracted by the opportunity to develop something cutting-edge, although they sometimes then miss the opportunity to solve equally critical but more-mundane problems.

Concept development and demonstrations are undertaken by NGOs, internal UN technology development teams, UN agencies in partnership with private companies and NGOs (e.g., the blockchain and biometrics deployment in Jordan), universities, innovation accelerators (e.g., the WFP Innovation Accelerator), organizations hosting hackathons, and sometimes volunteers conducting open-source development.

2. Product Development and Deployment

In this step, the concept is transformed into a product and scaled up as needed. This may mean taking the concept or prototype developed in the previous step and developing a technology product from scratch, adapting an existing product, or using an off-the-shelf product in a new or innovative way.

In the case of the WFP Innovation Accelerator initiative, which seeks to identify and cultivate solutions to hunger, WFP personnel assessed concepts for three to six months by developing a demonstration and assessing utility and impact. If a concept was promising, they obtained funding from donor governments to develop a full-scale product or adapt an existing one. Other UN-led initiatives work internally to bring a product or a technology to maturity and eventually deploy it to full-scale development. At the same time, interviewees also discussed instances in which NGOs struggle to enter the product deployment phase, mainly because of a lack of donor interest to fund full-scale development and deployment. Deployment refers not only to getting a technology product out in the field but also making the appropriate organizational and business process changes to allow it to reach its full potential. Sometimes solutions do not reach appropriate scale to become useful. An interviewee mentioned that there was "not enough coordination among humanitarian aid organizations so that they could form a customer base," and thus fiscal sustainability based on a fee or subscription model would be difficult at best. In other cases, the necessary organizational and procedural changes are not properly implemented, leading to inefficient uses of technology.

3. Content Development

Many tools come with some original content (such as databases or educational content), which then needs to be expanded and adapted to the needs of aid organizations or refugees. Content might be developed by users, volunteers, or refugees, or it might be outsourced to private companies or NGOs. Once a tool is developed and deployed, content development is easier to achieve, and we heard that help or input from volunteers or refugees can best be utilized at this stage.

4. Training

Successful deployment of a technology requires training for end users, which can include aid organization personnel, refugees, people responsible for data entry and content development, and maintainers and sustainers of the systems. For aid and nonprofit organizations developing and deploying digital technology, there are particular skills required to pursue digital transformation (NetHope Solutions Center, undated). Note the distinction between *digitization* (the technical process of converting analog information into digital form, such as taking a paper form and converting it to a digital format on a tablet or smartphone) and *digital transformation* (how individuals, businesses, or societies use digital data to fundamentally change a task or a process, and this includes not only the technical implementation but also the cultural and organizational changes) (Khan, 2017). Several interviewees expressed concern that training is often an afterthought and noted that it requires significant up-front planning and investment to maintain and sustain the training for the life cycle of the product. In the words of one interviewee, "The worst thing you can do is implement a software without a plan for training, uploading, sustaining, maintenance. . . . There's a lack of awareness of the amount of time it actually takes to do the procurement and to do the training."

5. System Sustainment and Maintenance

Although system sustainment and maintenance are related concepts, there are important differences. *Sustainment* is "the process, procedures, people, materiel and information required to support, maintain and operate the software aspects of the system" (Lapham, 2006). *Maintenance* is "the process of modifying a software system or component after delivery to correct faults, improve performances or other attributes, or adapt to a changed environment" (Institute of Electrical and Electronics Engineers, 1990, Std. 610.12).

Cost estimates of system sustainment and maintenance in humanitarian settings are not readily available, but these costs could average about half of the total cost of developing and deploying the system (Boehm, 1981). In the case of complex national security applications, such costs could be as much as 70 to 90 percent of the total life-cycle cost (Schmidt, 2011). We expect that these costs for technology deployed in humanitarian settings will be high (especially when security updates and patches are critical because of cyber threats and the vulnerability of refugee data) and will require careful consideration and budgeting up front. Many interviewees highlighted the issue. One said, "There is a lot of money for pilots. There is not a lot of money to maintain, and that is bad." And another explained, "So the challenge is: How do you capture that benefit, quantify it, create a system where it's worth the significant investment, training, and sustainment over time that it takes for that infrastructure?"

6. System Phaseout and Retirement

Planning how to retire or phase out a system is part of the normal system development life cycle. Although this phase is a well-known issue in the software development community, only one person in our interviews mentioned it, noting that it is important to assess when "to phase out a program based on need, because not all contexts require continuation of a tool, or a hand-over plan to a partner so that you can hand over the service being shared." Sustaining and maintaining software-intensive systems is expensive, and, as more of these systems are deployed in humanitarian contexts, the cost of sustaining them adds up while older systems become less productive. Retiring systems requires planning to ensure that users can be transitioned to a new system or platform. This process can be messy and expensive. Planning is needed to ensure a smooth and cost-effective transition.

Profit Formula

The concept of profit is often frowned upon in the context of humanitarian assistance; indeed, several of our interviewees expressed discomfort with the idea, while several others noted that this discomfort posed an unfortunate barrier to engaging more private-sector talent in addressing problems. Yet there still needs to be some kind of cost-benefit analysis that can show investors and shareholders that the benefits are worth the costs when developing technology for refugee contexts.

Costs follow from the activities and processes depicted in Figure 5.1, although initiatives do not always include all of those components. Interviews revealed several types of costs, such as traditional costs for concept development and prototyping, product development and deployment, project management, and software licensing fees. We also heard about the hidden costs of assessing needs and identifying requirements, training, deployment, sustainment, and maintenance. One cost that interviewees noted is frequently overlooked is power and internet connectivity at the individual level (the last mile). As one interviewee put it, stakeholders "should maybe focus a bit more on the program side in terms of how does that work in the field in terms of digital initiatives and projects and last mile and . . . reaching off the grid in hard-to-reach areas." In hard-to-reach areas, lack of connectivity may limit deploying a technology and should be factored in as part of the costs.

In refugee contexts, benefits may be direct (e.g., value created for refugees or aid agencies) or indirect (e.g., development of intellectual property that has value for other purposes). Although we heard about a range of ideas for deriving revenues, the majority of revenue streams to fund these technology initiatives appeared to originate primarily through grants from governments, private companies, foundations, and competitions. These entities often provide funding on an annual basis, which hinders maintenance and sustainment. To address this, one interviewee had the following suggestion:

> The first thing I would do is I would require multiyear [relationships]. . . . Most donors grant funds year-to-year, which creates massive instability. I only have one partner, one donor out of a dozen that has guaranteed their funding for more than one year . . . . Because tech companies move so quickly, . . . even with that partnership every year, we redesign the partnership. It is a challenge.

As interviewees described, some initiatives are able to quickly transition to a for-profit or nonprofit company with a subscription or fee-for-service model (or, more rarely, a pay-for-license model) that is able to scale up over the long term. In other cases, UN agencies or governments provide stable longer-term funding to maintain technologies. Other models focus on serving larger, more-affluent populations while supporting vulnerable populations as a small part of the business. Some approaches, such as cash cards, charge a fee per transaction. In some cases, a private organization retains ownership and copyright over a product and introduces it into other markets while the UN retains a free licensing agreement in perpetuity. In other cases, an organization monetizes data sets collected through the application.

Examples of indirect benefits include technology development and intellectual property with potential future use, such as IrisGuard's mobile biometric device (Soliman, 2016). Private companies can also engage in humanitarian initiatives to boost international brand development and marketing. For example, some defense and intelligence surveillance industries have sought "the legitimacy provided by partnerships with humanitarian actors" (Kaplan and Easton-Calabria, 2016). In addition, some companies join humanitarian projects to increase employee morale and job satisfaction, which, in turn, improves recruitment and retention. And in the internet economy, an organization might find that, when it develops a technology platform for refugee locations using donor funding, it becomes more feasible to offer a similar platform to the general population in adjacent areas when such an offering would not otherwise be profitable. Furthermore, doing so could establish the technology platform as a market leader.

Barriers to (and Facilitators of) Developing and Deploying Technology in Refugee Settings

Interviewees often described examples of barriers to the successful development and deployment of technologies in refugee contexts, as well as some facilitators of success. One of the most-common themes in many interviews was that a lack of a system-level approach to deploying technologies was a detriment to success. In this section, we describe the examples from the interviews, grouped by barrier. In many cases, the fact that one approach is a barrier makes clear that the alternative approach is a facilitator.

Short-term thinking. A short-term mindset was consistently identified as a key barrier to success. We learned about several examples of organizations focusing on the short-term demonstration of a technology rather than the longer-term deployment and sustainment of the services it provides. On the other hand, several examples of longer-term approaches were offered as success stories. Long-term success was felt to require planning for the entire life cycle of the technology, as well as development of the long-term partnerships and funding models needed to maintain the capabilities long enough for them to become relevant.

Projects driven by funding instead of needs. Donor priorities and available funding streams, rather than analysis of needs, disproportionally affect what projects get pursued. Interviewees felt that donors tend to favor short-term projects in regions in the news. They often compete with each other to fund multiple small efforts rather than developing consortia that will achieve scale and develop common technology ecosystems. On the other hand, several interviewees mentioned that it is difficult to assess needs within and across vulnerable populations and to measure the success of technologies deployed, although two interviewees did mention having a needs assessment methodology. Nonetheless, we learned about several cases in which a careful analysis led to successful projects that first focused on critical needs, then explored how to provide a service to meet those needs, and *then* pursued the necessary funding streams.

An emphasis on growth rather than on economies of scale. We heard from several interviewees that many initiatives measure success in terms of how many users they have or transactions they service. In order to grow, providers sometimes try to become a one-stop-shop for everything:

> You have some providers that say, we can do everything. We can give you an all-in-one solution and give you everything. The reality is, though, that these all-in-one solutions, some of which UN has developed, are not very agile and aren't necessarily taking in the new things that are coming online.

However, when it comes to software, scaling is different from growth. As one interviewee put it,

> The idea is you make a small investment now and get outsized returns. . . . But there's a reason that doesn't work in human rights and humanitarian aid. The problem is this: The way software delivers economies of scale, when you get over the initial investment, it gets cheaper to offer this to people. To say it gets cheaper on the margins doesn't make it cheaper; it just gets more economical.

In other words, scaling for humanitarian efforts is not measured purely in terms of the number of users but rather the ability to service more and more users with fewer and fewer resources.

For applications that do not require any special customization to address refugees' specific needs (e.g., social media or video conferencing), market forces select the most suitable technologies that refugees commonly use; for example, Facebook and Skype are free and fit refugee needs. But for vulnerable populations in crisis situations, free markets do not always produce efficient results (Levine, 2017). These markets tend to be underserved by few market actors, dominated by NGOs and government organizations without a profit motive. Because of this, we found that scaling could be impeded by two types of barriers. One involves different small initiatives that all try to solve the same problem, competing with each other for the user base, and none attaining efficiencies of scale. As a result, an organization might try to create an all-in-one solution, which could lead to growth but not efficiencies of scale. A more efficient approach would be creating partnerships to fill the gaps in services. A fragmented set of companies and NGOs competing and further fragmenting the customer base is the other significant barrier to scaling; many aid organizations do not reach a scale big enough to deploy efficient solutions and have difficulties absorbing multiple technical solutions at the same time.

Several interviewees highlighted the need for a technology ecosystem in which different companies and NGOs can innovate and provide interoperable solutions that together could achieve the economies of scale required. Others highlighted the need to group initiatives together to achieve efficiencies and scale. However, we heard few concrete examples of such efforts in this problem space. As one interviewee put it, "we have scarily few examples of tech that has scaled across systems and across orgs."

A focus on the technology rather than on changing business processes. Organizations often focus on the technology itself as a solution more than on the needs they are trying to service. But attaining efficiencies requires implementing organizational changes and fundamentally redesigning the way business is done (Brynjolfsson and Hitt, 1998). As one interviewee put it, "it should not be about individuals coming up with solutions. . . . Tech is not a gadget but needs to be a change in how we work." Another interviewee highlighted that, primarily because of funding limitations and lack of expertise, "nonprofit[s] . . . often tweak things in the margin instead of step[ping] back to say, how can we do this differently?"[2] We also heard several examples in which an organization accompanying the technology with a fundamental change in the way it did business was a facilitator of success. A few of the interviewees used the term *digital transformation* to describe these efforts. Digital transformation is not about the technology per se but rather about managing people, processes, and change across the entire system life cycle.

Regulatory and organizational complexities. With refugees in multiple countries and many aid organizations involved, it is more difficult for organizations, especially in private industry, to deploy solutions across geographical, linguistic, regulatory, and complex organizational barriers. In general, private industry participants want to quantify risk regarding any initiative they consider. Several interviewees mentioned private industry's reluctance to work in certain countries or to enter into partnerships with certain government agencies. In some countries, estimating risk is difficult, given the volatilities and the complexities related to refugee populations. The challenge is compounded by organizations having less authority to make decisions on how to imple-

---

[2] This characterization might apply to some NGOs, but we also heard several examples of NGOs pursuing transformational changes.

ment innovations because innovations must be coordinated within a larger web of UN and government agencies and NGOs that change over time and across regions. These circumstances lead to an aversion to take risks. As one interviewee put it, "People don't take enough risks in humanitarian space, but we don't get enough risk and patience and willingness to put impact before profit." On the other hand, another interviewee noted, "To introduce new technology, you have to take risks."

Lack of a system-level approach. Many of the barriers already described are associated with a broader cross-cutting theme that emerged across several interviews: the lack of a system-level approach to thinking about technology in the context of refugee and humanitarian aid. In the words of one interviewee,

> There have sprouted a lot of fragmented efforts and a lot of piecemeal efforts with companies or other actors trying to help where they can and to be useful, productive, and impactful, but the more-systemic issues just haven't really been addressed.

Another interviewee focused on the aid system struggling to keep the big picture in mind:

> Humanitarian agencies broadly are quite good at tactical creative problem-solving. . . . I think as the aid system has grown and become more professionalized, they've become more bureaucratic and less good at systematic problem-solving or a higher level of innovation, which is: What *should* we do rather than *how* do we do what we do better?

Another interviewee distinguished between innovating *within an organization* versus systematically promoting changes *across organizations*:

> It depends what you mean by systematic. [Some] organizations are good at doing innovations systematically—so, doing innovation repeatedly and bringing up new things. And they are . . . quite good at making internal changes based on those solutions. But I think everyone struggles with getting changes outside their organization.

This lack of clear system-level planning and execution makes it difficult for private-sector participants and NGOs to understand how they fit into the broader system and come up with a business model that is viable and provides value. As one interviewee put it,

> We know there are so many stakeholders in the humanitarian and development sphere/landscape/market that would derive some benefit from digitizing operations. So, how do we then make a business model for the delivery of these solutions? How do we make it appealing, compelling, fair to those implementers?

Several interviewees described how private-sector participants struggle to find their place in this context, and, as one interviewee put it, they probably need to think about planning differently: "You can't go fully in the business mode; you can't go fully in the humanitarian mode."

Tools for Applying System-Level Thinking to Support the Development and Deployment of Technology in Refugee Contexts

Although a lack of system-level thinking is a challenge for organizations working to develop and deploy technologies in refugee contexts, there are tools available to help build a more systematic approach.

Questions to Ask to Guide a System-Level Approach to Developing and Deploying Technology in a Refugee Setting

We provide one such tool in Table 5.2, which outlines important questions to ask to guide a system-level approach to developing and deploying technology in a refugee setting so that the approach considers the entire ecosystem rather than individual uses, applications, aid agencies, time periods, and so forth. We group the questions by system focus, as follows:

- *Refugee technology needs and priorities.* UN agencies, governments, aid agencies, NGOs, and private companies can come together to prioritize needs and develop

Table 5.2
Questions to Ask to Guide a System-Level Approach to Developing and Deploying Technology in Refugee Settings

System Focus	Questions to Ask to Guide the Approach
Refugee technology needs and priorities	What are the needs? How should they be prioritized? How can all the relevant organizations (i.e., UN agencies, governments, aid agencies, NGOs, private companies) work together to maximize value across refugee populations and countries?
Legal, cultural, and other issues relevant to the context of the specific refugee population	How does a given technology apply in the specific country context (e.g., legal, cultural)? How should technology opportunities be framed for the context to serve pressing needs while maximizing benefits and minimizing risk?
Coordination of private-sector, government, and NGO efforts to meet specific needs	How can private companies, governments, and NGOs coordinate their efforts to serve a specific set of refugee population needs? How do the organizations' business models intersect and influence or reinforce one another?
Development of business models to help private companies and NGOs participate	What new business models are needed to help other private companies and NGOs participate in technology initiatives for refugee populations? How can the needed elements be put together systematically to maximize value while minimizing risk?

a framework for working together to deploy technologies efficiently while maximizing value across multiple refugee populations and aid agencies in multiple countries.

- *Legal, cultural, and other issues relevant to the context of the specific refugee population.* Stakeholders can systematically evaluate how a technology applies to a specific refugee context. Some issues are dependent on the local legal framework or cultural context to understand how to frame a set of technology opportunities.
- *Coordination of private-sector, government, and NGO efforts to meet specific needs.* Stakeholders can consider the joint perspective of the relevant set of private companies, governments, and NGOs and how they can work together as a system to address a specific set of refugee population needs.
- *Development of business models to help private companies and NGOs participate.* Stakeholders can require a standard framework for developing an appropriate business model to help private companies and NGOs engage in technology initiatives (see Chapter Seven for further discussion). The traditional for-profit business model applies but does not cover all potential opportunities. The business model components described earlier in this chapter can be put together systematically to help stakeholders understand the relationship between parts of the model and how the stakeholders focus their efforts. A successful model should take into account systemwide interactions in order to minimize risks while maximizing value.

System-Level Approach for Evaluating the Application of Technology in Refugee Settings

While Table 5.2 offers a way to ask more-systematic questions about how technologies might be deployed in a refugee context, we also sketch out a broad system-level approach for evaluating the application of technology in a refugee setting. This approach was inspired by Michael Porter's five forces of industry (Porter, 1979),[3] which private companies use to evaluate a specific competitive environment or a given industry and assess the potential for profitability, level of risk for entering that market, and strategic value of pursuing that opportunity. Instead of focusing on a market, our approach focuses on the refugees as a vulnerable population and assesses five drivers that influence how effective the application will be (Figure 5.2). The five drivers, selected based on interviewee comments, are as follows:

- *Region of origin.* Ongoing political and security considerations associated with the region of origin may influence whether a technology could apply and how to approach it for many years after the migration.

---

[3] Porter analyzes five forces that potentially affect profitability: existing industry rivalry, threat of new entrants, threat of substitute products or services, bargaining power of suppliers, and bargaining power of customers.

Figure 5.2
Five Drivers for Evaluating the Application of a Technology Solution in Refugee Settings

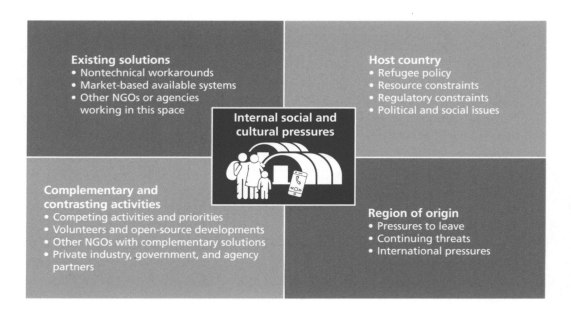

- *Host country.* Local regulations or attitudes in host countries affect the potential of a technology solution. Assessments of technology in a specific context need to take local realities into account.
- *Existing solutions.* For every current problem, there is some current method for handling it, even if there might be a more efficient way of doing so. For example, existing solutions might be applied to the same problem in different regions or contexts. And other organizations often are pursuing various solutions to the same problem. How effective a given technology's application is should be considered in comparison with existing options.
- *Complementary and contrasting activities.* There is rarely a single organization that can be everything to everyone. Partnerships could be a determinant of success for any given technology, but successful partnerships depend on compromise.
- *Internal social and cultural pressures.* Different populations' language skills, technology savvy, level of education, attitudes towards gender or race, and attitudes to privacy or security can affect the success of technical solutions.

The region of origin, the host country, and the internal social and cultural pressures set the population's sociopolitical context and associated risks and constraints for applying any technology. The existing solutions and the complementary and contrasting activities describe the competitive and collaborative environment that could translate to risks and opportunities. Note that this is a high-level framework, and the factors vary across refugee settings that would need to be assessed anew in each technology development and deployment situation.

Conclusion

This chapter described business models involved in developing technology solutions in refugee settings. There are four main components of business models as applied in this context: value proposition, key resources, key processes, and profit formula.

In addition, we outlined several barriers and facilitators that can influence the successful deployment of technology in refugee settings. Barriers include a short-term mindset, projects driven by funding rather than needs, an emphasis on growth rather than on economies of scale, a focus on technology alone while missing the opportunity presented by changing business processes, and regulatory and organizational complexities. In addition, an overarching barrier highlighted in many of our interviews is the lack of a system-level approach to thinking about technology in the context of refugee and humanitarian aid.

Finally, we outlined available tools to build a more systematic approach to technology deployment in refugee contexts. We described several questions that can be asked at each stage of the process and then presented an approach for evaluating the application of technology in a refugee setting and what may determine whether that application is a success.

CHAPTER SIX

Ethical, Security, and Privacy Issues Related to the Use of Technology in Refugee Settings

As indicated in previous chapters, technology can be used to benefit refugees, aid agencies, and other entities involved in refugee settings. But increased use of data raises ethical, security, and privacy issues, which are the focus of this chapter. These issues encompass four interconnected areas of concern: (1) the humanitarian sector's lack of thorough, shared ethical frameworks and safeguards to address technology risks; (2) data responsibility, including data protection and rights; (3) technology's potential to introduce or exacerbate bias; and (4) conflicts of interest.

Certainly, discussions of the hazards of emerging technologies often raise concerns in other contexts. However, the refugee context complicates the conversation about technology and ethics because refugees have fled violence, persecution, and other dangers for often precarious new circumstances.

At the same time, technology solutions can enhance protections, safety, and legal rights in refugee situations to the extent that *not* using technology can yield yet greater risks. Indeed, in 2016, the UN included internet connectivity as a human right in the Universal Declaration of Human Rights (Howell and West, 2016). Thus, ethical humanitarian use of technology involves balancing technology's value against its possible misuses and evaluating and mitigating its risks. In this chapter, we discuss those considerations in each of the four key areas of concern.

Ethical Frameworks and Safeguards to Address Risks

The humanitarian sector lacks comprehensive, common frameworks to safeguard ethics, security, and privacy in relation to technology and data use (Raymond et al., 2016). This is, in part, because the role of technology and data in aid provision has expanded rapidly, preceding rather than following the development of shared humanitarian standards (Coppi and Fast, 2019; Latonero et al., 2019; Raymond and Harrity, 2016). As one of our interviewees stated, "In most areas where we try to push tech, it's totally uncharted territory." Still, there are some existing starting points that the sector could build upon to work toward such standards. This section summarizes the current landscape, as well as challenges inherent in establishing standard guidelines for the ethical and responsible use of technology and data in refugee contexts.

Foundational Ethical Frameworks and Existing Efforts

For humanitarian work in general, globally accepted ethical frameworks and standards already exist (Raymond and Harrity, 2016). The principles of humanity, neutrality, impartiality, and independence were adopted by the United Nations General Assembly and underlie the work of many humanitarian organizations (Capgemini Consulting, 2019; UNHCR, 2015a). Other principles, such as "do no harm," are also widely recognized (Raymond, 2017), and several documents have been developed through consultative processes and have achieved multi-organizational commitment. Examples include *The Core Humanitarian Standard on Quality and Accountability* (CHS Alliance, Group URD, and the Sphere Project, 2014), the *Code of Conduct for the International Red Cross and Red Crescent Movement and Non-Governmental Organizations (NGOs) in Disaster Relief* (International Federation of Red Cross and Red Crescent Societies and International Committee of the Red Cross, undated), and *The Sphere Handbook 2018: Humanitarian Charter and Minimum Standards in Humanitarian Response* (Sphere Association, 2018). These standards are, of course, still applicable when technology is involved.

Existing frameworks for technological innovation and responsible data use in humanitarian and development settings are more fragmented. Many individual organizations have their own internal policies, principles, and guidelines (Capgemini Consulting, 2019; Raymond and Harrity, 2016). For example, the UN adopted the *Guidelines for the Regulation of Computerized Personal Data Files* in 1990 (United Nations General Assembly, 1990; Raymond et al., 2016). UNHCR published a "Policy on the Protection of Personal Data of Persons of Concern to UNHCR," which provides principles for personal data processing, such as necessity and proportionality (UNHCR, 2015b). A UNOCHA report contains a framework for analyzing ethical principles in humanitarian innovation (Betts and Bloom, 2014), which charts innovation principles at the individual level, community level, and system level. Another UNOCHA document describes minimum standards for humanitarian data use (Raymond et al., 2016). In addition, the International Committee of the Red Cross published a *Handbook on Data Protection in Humanitarian Action* (Kuner and Marelli, 2017). A Privacy International and International Committee of the Red Cross report lists data-related questions that organizations should consider to prevent harm in cases of technology use (Kuner and Marelli, 2017). The Harvard Humanitarian Initiative's *The Signal Code: Ethical Obligations for Humanitarian Information Activities* stipulates data-related rights of people in crisis contexts and obligations for humanitarians (Campo et al., 2018). And "The Blockchain Ethical Design Framework," a 2019 paper, contains a decision tool to assess whether blockchain is an appropriate technology choice (Lapointe and Fishbane, 2019). Similarly, some of our interviewees described data protection policies at their organizations. However, others noted the need for more overarching frameworks in this area.

Other efforts have sought and achieved endorsement by other groups. For example, the World Bank's 2017 *Principles on Identification for Sustainable Development: Toward the Digital Age* advocates inclusion, design, and governance principles and is endorsed by a variety of humanitarian organizations (World Bank, 2017). Multiple organizations have endorsed the principles of donor alignment for digital health (Digital Investment Principles, undated). The Digital Impact Alliance's principles for digital development include nine "living" guidelines for development actors to use when designing and enacting technology-related projects (Principles for Digital Development, undated). And discussions involving more than 100 organizations led to a document about implementing those principles in practice (Waughman, 2016).

Remaining Gaps and Barriers

Although some of the noted frameworks have pursued broader consensus, they did not, for the most part, result from wide-reaching, organized, and inclusive deliberation processes and have not achieved international, multi-organizational endorsement and commitment. Several barriers make such a systemwide result difficult to accomplish. First, humanitarian actors lack mutual consensus on the risks to vulnerable groups that stem from technology and data-related efforts (Raymond and Harrity, 2016). Second, these actors may lack knowledge of related legal and regulatory requirements; for example, many European nonprofit organizations are unaware of or misunderstand the General Data Protection Regulation, which governs data privacy for the European Union (Capgemini Consulting, 2019; Coppi and Fast, 2019). Third, there is no consensus about which technologies would need minimum standards (Raymond and Harrity, 2016). Emerging technologies, such as digital ledger technologies and new digital identity systems, are still hypothetical or in early stages of deployment in humanitarian contexts (Coppi and Fast, 2019; Juskalian, 2018; Mercy Corps, 2017). Fourth, technology and data use in humanitarian settings involve many actors, as described in Chapter Two, and each has different perspectives and roles (Raymond et al., 2016). It is challenging to create holistic guidelines that incorporate all who are operating in such a dynamic space. Finally, agreement and balance between broadness and practicality are difficult to achieve. On the one hand, to be all-encompassing, attempts at sectorwide guidelines could be too broad to translate into concrete application (Latonero et al., 2019; Schiemichen, 2018). As one of our interviewees pointed out, there is a big jump between ethical principles and practice. On the other hand, proposed guidelines may include many specifics that do not apply in certain contexts. It is perhaps for this reason that the existing examples described earlier either are quite general or are organization-specific. Addressing the gap in the middle ground requires evaluating risks that arise in various situations across humanitarian technology efforts in the areas of data responsibility, bias, and motivation.

Data Responsibility

Humanitarian operations prompt the collection, creation, use, sharing, and storage of increasingly vast amounts of digital data about refugees, aid providers, and operations. Digitization and remote storage of data can, in some cases, serve as a security improvement over local storage (USAID, undated). However, at each stage of data processing, risks of data disclosure, misuse, and error increase the need for and responsibilities inherent in data responsibility. Specifically, data responsibility includes protecting data from threats to privacy and security and preserving data subjects' rights to informed consent and data correction.

Data Protection

Cybersecurity problems and data-sharing practices can expose refugees' personal data to access by entities other than the original data collector. Those with access to refugees' data could misuse it in ways that cause harm. The primary privacy and security threats to refugees are scams and surveillance (with varying risks, depending on the motivation of the actor).

Among aid providers and refugees alike, levels of digital literacy, online security awareness, and related training vary widely (Latonero et al., 2019; Pirlot de Corbion et al., 2018; Simko et al., 2018). Aid groups may lack the ability to protect beneficiary data or the knowledge of how to do so. Illustrating this, one study describes aid offices using an unsecured Wi-Fi network, transmitting data via an unencrypted website, and storing sensitive data in the cloud (Latonero et al., 2019). One of our interviewees expressed concerns about some organizations' use of online interpreters who translate private information about refugees' health, children, behavior, and marital relations without the organizations really knowing whether those interpreters have been sufficiently vetted. Furthermore, aid groups may not be aware of what information third parties are collecting in relation to their work (Pirlot de Corbion et al., 2018), which can put beneficiaries at risk. Metadata, such as call detail records, may enable aid groups and malicious actors alike to better understand population movements (Raymond et al., 2016).

Meanwhile, some aid providers take precautions to protect data. Several interviewees described their organizations' efforts, such as using secure servers. One described an organizational call center that allows refugees who do not feel safe providing personal information in on-the-ground data collection scenarios to do so privately. Provision of personal mobile devices or internet connectivity to refugees can itself enhance their privacy and data security. As one of our interviewees explained, before mobile phones, refugees had to wait in lines to have phone conversations in front of others. Now they can have confidential conversations. Another interviewee's organization in Latin America uses internet-accessible interactive service maps for individual users rather than social media to disseminate information, "because people don't want to be tracked and it can be quite dangerous to have people posting questions and information like they do in Europe." The interviewee further stated, "We do take down information that's posted about people individually or things that could lead people to them, like phone numbers or addresses. We don't hold onto people's data."

Refugees' knowledge of security and data protection methods likewise varies. Refugees may not be familiar with such practices as secure password creation (Simko et al., 2018). One of our NGO interviewees stated, "We create the clients' passwords and we make them very simple, so they are probably easily steal-able." This interviewee also indicated that scammers have capitalized on some refugees' lack of awareness of their data's vulnerability by targeting these refugees for financial or identity theft scams: "They are very vulnerable in general, so I'm guessing they could fall into any trap." Another reiterated that some refugees "don't have the same awareness of how the information could be used against them, so behavior is different." Just like any other group, refugees may not be able to distinguish online misinformation; however, in refugees' case, targeted misinformation could expose them to physical danger and other harms while in transit and afterward (Benton, 2019).

Then again, some refugees are very conscious of online security measures. One interviewee noted that refugees can be particularly adept in protecting their own data: "Every average Syrian is very in tune with all the proxies available in the market. . . . Even those who are illiterate know about proxies and [virtual private networks]." Encrypted messaging apps, such as WhatsApp and Telegram, have been popular among Syrian refugees. Our interviewee explained that Syrian refugees thought the Telegram app was especially secure because it was Russian, it allows the ability to follow channels without having to "be friends," and it is more tied to the phone than the internet. Although refugees are largely using common U.S. social media apps, we heard examples like this showing that the refugees have some fear of surveillance from multiple angles.

Surveillance and the harms that come from it are what aid groups and refugees alike see as a primary threat motivating cybersecurity practices and workarounds. Refugees and their families may face persecution, reprisals, xenophobia, and stigmatization (Latonero et al., 2019; UNESCO, 2018). They fear surveillance from the groups behind such harms, which can include the government of the country they have fled, other groups back home, foreign governments and intelligence agencies, and host country governments. Ranking what Syrians consider to be the main security threats online, one interviewee stated,

The Syrian regime is probably number one. Or anyone who could share [their data] or exploit it with the Syrian government. Number two is, some people are concerned about the U.S. or Russia or other countries involved on the ground that could also spy. Because of the situation, there's a lot of revenge and people telling on others. Anyone could go and stick a tracking device on somebody and say this guy is [with the Islamic State] and so forth.

Similarly, refugees are sometimes skeptical of electronic equipment, such as smart-phones and laptops, distributed by aid groups. Another interviewee explained, "Refugees show caution and anxiety about technology used by Western humanitarian organizations. They worry data may be used for intelligence purposes." Another linked these fears with refugees' wariness of providing their data in their new location, explaining that refugees from Myanmar in Bangladesh were reluctant to provide their personal data because the government of Myanmar had used such data for harm.

Data-Sharing Among Aid Agencies

There are debates about the pros and cons of data-sharing and interoperability among systems that contain refugee data, as well as the conflicting values and goals involved. On the positive side, interoperability could increase the efficiency of humanitarian programming. For example, it could allow refugee health information to be accessed by different health care aid providers even if refugees move, ensuring that refugees receive consistent care (Latonero et al., 2019). It could also help lessen the fatigue that refugees experience from repeatedly having to provide their information and recount their hardships to uncoordinated aid providers. On the negative side, if aid agencies transmit data to third parties with weak protection standards or differing motivations, it could expose refugees to security risks (Nonnecke, 2017). Interoperability could also enable "function creep," whereby data are used for purposes beyond the original intent (Soliman, 2016).

Biometric data are especially sensitive. On the one hand, host countries have a responsibility to maintain security and enforce their laws; they may seek to search refugees' biometric data for counterterrorism or law enforcement reasons. However, this also means that personal information provided by refugees could be used to police them (Schiemichen, 2018; Soliman, 2016). For example, a 2013 update broadened the mandate of the European Asylum Dactyloscopy Database—which collects fingerprint data from asylum seekers—to allow Europol access to the data (Latonero and Kift, 2018; Orav, 2017). European Union policy prescribes that people arriving in Europe should apply for asylum in the first country they enter; thus, refugees who are residing outside their country of arrival fear being sent back to that country (frequently Greece or Italy) by police using the fingerprint data for identification purposes (Latonero et al., 2019).

The motivations of those who hold data can also change over time, leading to risks from data-sharing and interoperability in politically fraught situations (Latonero and Kift, 2018). Interviewees talked about the ethical conundrums faced by aid agencies in balancing the security interests of refugees against those of the nation-states that host them. Dangers can result if governments use the data to target particular groups for harm (Soliman, 2016). One of our interviewees questioned the sharing of humanitarian data with governments: "What if, in the future, that nation-state is not friendly to that group coming in?" Another interviewee's organization will not share sensitive data that could be abused, such as ethnicity information. Similarly, both host countries and source countries may eventually want to use data for repatriation purposes, regardless of whether refugees feel safe returning home. When the Bangladeshi government and UNHCR were conducting biometric registration of Rohingya refugees, some worried that the effort could eventually be used for forced repatriation, a practice not legal under customary international law (Rahman, 2017). Finally, centralized databases like those of the UN and Europol can be a particularly desirable target for breaches (Schiemichen, 2018; USAID, undated).

Improper data protection can also make refugees reluctant to provide their data at all or trust humanitarian aid in general. Refugees' anxieties about surveillance can cause them to avoid humanitarian structures and forgo access to rights and services (Latonero et al., 2019; Raymond et al., 2016). Fears of privacy and security threats could also lead refugees to give false personal data to aid providers, undermining the providers' ability to serve the refugees and others (USAID, undated).

Data-Related Rights

Considerations of data-related rights in refugee settings include consent, agency over data, and data correction.

In considering the matter of consent, stakeholders must examine whether refugees are in a position to give voluntary, informed consent to participate in technology-enabled aid initiatives. One of our interviewees framed the issue in this way: "People are showing up and fleeing violence and war and are being asked to give information about themselves to an agency in order to receive basic life-saving care and are not being told what is being done with that data." In other words, refugees must provide often sensitive personal information as a condition for receiving aid (Latonero et al., 2019). Even if aid organizations provide refugees with informed consent protocols, these do not always convey information in an understandable way or empower refugees to decline (Latonero et al., 2019). Additionally, in humanitarian crises, the sheer number of refugees in need of assistance may overwhelm aid providers, overshadowing consent-related concerns.

On the other hand, as refugees cross a border into another country, the host country has a right to collect information about those who have crossed (sometimes relying on UN agencies as the data collectors), just as those countries would collect information about other people crossing their borders. Furthermore, combatants from the same war the refugees are fleeing could try to blend in with the refugee populations, which happened when civilians were fleeing the Islamic State of Iraq and Syria and the military operations against the group in Iraq, for example (Culbertson and Robinson, 2017).

Other considerations for data-related rights pertain to refugees' agency over their data (Capgemini Consulting, 2019). A particular concern is whether refugees can opt out of humanitarian data storage mechanisms and transition into more long-term, mainstream options (Schiemichen, 2018). Once refugees are settled and integrated into a new society, they may not want aid organizations with whom they no longer interact to continue to possess their personal data. There is also a question of how long agencies should retain personal data collected from people of young ages (Latonero and Kift, 2018). Storing children's data can subject them to difficulties transitioning from aid systems to mainstream ones. On the other hand, such tracking can be important to children's safety. Europol reported in 2016 that at least 10,000 migrant and refugee children in Europe had gone missing over two years; some were likely criminally exploited (European Parliament, 2018; "Over 10,000 Migrant Children Missing: Europol," 2016).

These concerns notwithstanding, one of our interviewees believed that such consent-related concerns are a red herring: "For someone who has just crossed the border who has nothing—do they care about privacy? I don't know. And I think they would rather prefer to waive any right to privacy in order for their kids to get vaccinated, or in order for them to have a roof over their heads." So, although some saw these trade-offs as a problem, others saw them as a fact of life. Outside the refugee context, in order to access a desired or required service, people commonly provide personal data when they might prefer not to do so, or they consent to terms and conditions without reading or understanding them. However, because of refugees' particular vulnerabilities, these situations are not entirely parallel.

Finally, mistakes in collected data (such as misspelled names) or missing data can cause problems for refugees, and the number of organizations collecting refugees' data for different systems has increased the chance and impact of data discrepancies. For example, researchers in one study learned of a case in which parents were threatened that their child could be taken away because of an inconsistency in their last names (Latonero et al., 2019). Compounding this problem, computers and automated systems create rigid rules that can then be inflexible in cases of inaccuracy. Without built-in mechanisms for data subjects to seek redress or correction of errors, they can be subject to such harms as inconvenience, emotional distress, and even security threats.

Bias

In addition to concerns about data-related rights and a lack of established safeguards, technology in refugee settings may introduce or exacerbate bias. Bias-related risks include the exclusion of certain groups from technology-based aid initiatives, as well as the reinforcement of inequality and discrimination via bias embedded in aid mechanisms and in automated data analysis and decisionmaking systems.

Exclusion

Access to technology is unequally distributed, so technology-based aid provision can exclude those who cannot use technology (Dette and Steets, 2016). As one of our interviewees pointed out, it matters "what tools people have access to and who doesn't have access to those tools." Use and possession of technology can vary by gender, age, and other factors (Latonero et al., 2019). One of our interviewees highlighted that there is no sufficient answer to the question, "What if people are mobility impaired, or with disabilities? How do they access these mechanisms?" And biometric systems may exclude those who lack readable fingerprints, hindering refugees from receiving aid through cash assistance and other mechanisms (Latonero et al., 2019). Moreover, when data are collected from populations to inform aid provision, such data can be nonrepresentative if those without access to technology cannot provide their input (Latonero et al., 2019).

Reinforcement of Inequality and Discrimination

Technological implementations also perpetuate aid providers' and host societies' conscious and unconscious biases (Frey and Gatzweiler, 2018). As a UNOCHA report states, "Social, economic and cultural biases in the way data is generated, collected, processed and analysed can lead to oversights and assumptions that further embed social and economic inequalities within affected communities" (Raymond et al., 2016, p. 5).

Refugees' histories of interaction with aid agencies can create digital trails and classification systems that expose them to profiling and discrimination (Latonero et al., 2019). One of our interviewees stated, "This person's identity will be tied to the fact that they are a refugee, and then there are biases or predispositions that will be latched on top of that." Such discrimination could occur if, for example, digital records of refugees' interactions with humanitarian agencies, such as participation in a cash transfer program, cause financial service providers and advertisers to profile them (Pirlot de Corbion et al., 2018), designating them ineligible to borrow money or singling them out for high interest rates. Conversely, another interviewee noted that being classified as a refugee has some benefits also, such as the legal protections that such status affords.

Growing artificial intelligence capabilities also pose privacy and security issues for refugees via big data analysis. Artificial intelligence technologies can infer individuals' characteristics, such as ethnicity and gender, from patterns and correlations in data sets (Molnar and Gill, 2018). Identity data in aggregation comprise demographically identifiable information that can enable the tracking of groups (Latonero et al., 2019). Such technologies are progressing more quickly than are mitigations of their flaws (Coppi and Fast, 2019; Osoba, 2018; RAND Corporation, undated), and they are subject to the same biases that are present in the data they are trained on. Refugees can be harmed by decision outputs of automated systems and algorithms. For example, certain countries use algorithms to help evaluate immigration and visa applications, increasing efficiency and reducing backlog—to refugees' benefit. However, the predictive analytics used to conduct risk assessments may incorporate embedded or purposeful biases that discriminate against those of particular religions or those from certain countries of origin (Molnar and Gill, 2018). Similar concerns were raised when a 2018 article in *Science* proposed an algorithm to assign refugees to resettlement locations based on historical data about employment prospects (Bansak et al., 2018) instead of refugee and community preferences or needs, such as health problems.

Conflicts of Interest

Technology solutions in refugee settings have multiple uses, including aiding refugees, supporting aid agency implementation, and helping governments manage security. Clear communication by organizations about their projects' intended results is important for evaluating whether envisioned results are proportional to the risks involved. When examining the motivations and interests in a technology initiative, questions should be asked about who the intended beneficiaries are, whether these beneficiaries' best interests are served with the initiative, what other interests may be drivers of the initiative, and whether implementation of experimental technology presents risks that outweigh the benefits. Answering these questions should help manage conflicts of interest.

Some interviewees expressed concern that the aid community's desire to be innovative led to implementation of unproven technologies. One explained,

> everyone . . . wants to say they are using the most cutting-edge tech. . . . All of the agencies are now fighting and competing to . . . pretend they know what blockchain is and position themselves to be able to utilize these technologies so that they are not stick-in-the-mud, old-school organizations that can't get with the times.

Another interviewee summarized, "Some see the innovation agenda as a resource mobilization opportunity rather than as an opportunity to be very self-critical." Additional sources in the literature discuss the harmful nature of a "technology-as-savior" mentality or "disaster experimentation" characteristic of startup culture that promotes a top-down approach to solving humanitarian problems (Coppi and Fast, 2019; Raymond and Harrity, 2016; Scott, 2016). One article echoed a sentiment heard in our interviews: "Humanitarian actors are, in many cases, deploying [information and communication technology] solutions in search of potential problems to solve, rather than first identifying the most urgent problems and then ensuring that the proper tool is being used correctly to address them" (Raymond and Harrity, 2016, p. 11).

These depictions from interviewees and the literature indicate that some technology initiatives pursue private-sector or organizational goals rather than solely the best interests of the refugees. For instance, private companies could try to influence technology choices to benefit their own tech products, or aid agencies may prioritize operating efficiency over refugees' well-being. Although it is natural for organizations and individuals to have their own internal goals and interests, the interests of the refugees should remain front and center in any technology initiative. Any organizational or personal conflict of interest should be clearly stated, addressed, and adjudicated with this core principle in mind. One of our interviewees noted,

> It's about how you do it, not what you do. . . . Every kind of tech can be used and piloted for humanitarian action if the protection frameworks are in place and [it happens] in a simulated, secure environment without exposing beneficiaries to any kind of danger—why not? . . . It's part of the risk that comes with innovation.

Conclusion

Ethical use of technology involves balancing technology's risks and benefits and managing security and privacy considerations. In this chapter, we described four key areas of concern:

- *Frameworks and safeguards* to address technology risks are underdeveloped and fragmented across the humanitarian sector, although some foundations exist.
- *Data responsibility* issues—including protecting data from misuse and respecting refugees' data-related rights—are growing more urgent and complex as aid agencies collect increasing amounts of personal data.
- *Bias* is introduced or exacerbated by technology-based humanitarian efforts when they exclude certain groups or perpetuate inequality or discrimination.
- *Conflicts of interest* often arise with technology initiatives in refugee contexts. Motivations for becoming involved in such an initiative can include benefiting refugees, improving operations of aid groups, and testing a new technology to meet organizational or personal objectives unrelated to the best interests of the refugees. Clarity about motivations, interests, and intended results is important in order to weigh risk and manage conflicts of interest.

When humanitarian organizations fail to account for potential problems in these four areas, technological initiatives can subject refugees to discrimination and risks related to security, privacy, and technology experimentation. Ethical, security, and privacy challenges have direct consequences for both refugees' well-being and overall levels of trust in the humanitarian system.

CHAPTER SEVEN

Conclusions and Recommendations

In this study, we considered how technology is used by refugees and those who help them. In particular, we described the roles and responsibilities of actors in this ecosystem; various uses of technologies in this setting; the ways in which refugees perceive technology; the business models through which such technology is developed; and ethical, security, and privacy issues. We found some broad, overarching themes. First, there are multiple actors with complex and interdependent relationships, and technology is changing their roles and responsibilities over time—for example, by creating new roles and simplifying or altering long-standing roles. Yet there should be better coordination of investment in, and use of, technology in refugee settings, which may provide more opportunities for private-sector engagement. Second, although most refugees and aid agencies rely mainly on mainstream technology applications developed for more-general audiences, there has been sizable investment in creating applications specific to refugee settings, most of which seem to fizzle out over time. Third, investment in technology in refugee settings is often made without preparing for the full system development life cycle, from project initiation to system retirement. Finally, technology in humanitarian settings is being implemented in advance of needed ethical, security, and privacy frameworks. It is with these themes in mind that we offer the following recommendations for stakeholders to develop and deploy technology efficiently, effectively, and ethically.

Focus Private- and Humanitarian-Sector Technology Investments More Strategically, Weighing Risks and Benefits and Considering the Full Technology Life Cycle

As documented in Chapter Three, there have been multiple investments in technology in refugee settings, many of which are not maintained or do not account for the full technology life cycle. Instead of fragmented investments in technology, there should be a more thoughtful and strategic approach to decisions about which projects to invest in. As described in Chapter Five, an often-stated challenge to investing in technology strategically was the lack of and need for a tool or framework to assess humanitarian technology investments both from the business perspective and from a broad system-level view within the sociopolitical context.

In the corporate world, one framework for capturing or creating a business model is the *business model canvas* approach (Osterwalder and Pigneur, 2010). This approach often means documenting the different elements of a business model (value proposition, key resources, key processes, and profit formula) and exploring how each element relates to others. The approach can be used to understand the systemic effect of specific choices on the elements of the business model. Researchers have modified the business model canvas for the humanitarian problem space (Blank, 2017; Gray et al., 2019) and for joint partnership business models (Dimarogonas, 2012).

We built on these models to propose a business model canvas tool tailored specifically to the application of technology in refugee settings and that a range of stakeholders, including aid agencies, private-sector companies, and donors, should use when determining technology investments in humanitarian settings. With this tool, shown in Figure 7.1, we aim to enable the evaluation of technology investments by considering actual needs; accounting for political, legal, cultural, and geographical barriers across countries; balancing risks and costs with derived benefits; accounting for the different stages of the technology life cycle; addressing risks and costs; and considering such barriers as digital literacy and last-mile internet and power connectivity.

Figure 7.1
A Business Model Canvas for Technology Investment in Refugee Settings

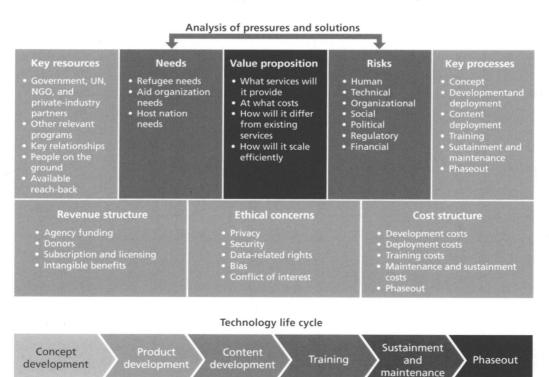

The first step in our business model canvas tool is to conduct an analysis of the pressures and solutions involved (see Figure 5.2 in Chapter Five) to understand context (needs, risks, key resources, the competitive or collaborative environment, associated opportunities or threats, and potential ethical concerns). Then, drawing on the results of that analysis, the second step is to build a value proposition of how the technology would meet needs and address risks and concerns while leveraging key resources available, address the key processes of the life cycle, estimate costs across the life cycle, and identify revenue streams to cover the costs. The analysis of pressures and solutions feeds into analysis of the needs and risks in the business model canvas. Stakeholders can then review interconnections among these various elements when developing strategies. We envision that these business model canvases can exist at different levels of abstraction, from strategic levels through detailed implementation. Different stakeholders should use this tool in different ways:

- *Donors (including governments and foundations) and aid agencies* should use this tool when developing their own strategic plans for technology investments and require their implementing partners to demonstrate that they have considered and planned for all of these issues when investing in new technologies. Much as a theory of change or logic model is used in international development and other programs to define the logical steps of a strategy (W. K. Kellogg Foundation, 2004), our proposed business model canvas can be used instead in settings that blend humanitarian and business interests to lay out logical steps and interconnections among goals and assumptions. Specifically, donors have the capability to lead strategic thinking on technology investments through what they fund; when giving funding for projects, donors and aid agencies should require the analysis that we propose instead of or in addition to the usual logic model approach to demonstrate that these issues have been thought out end to end. If large donors took the same approach, this could introduce common approaches to technology strategy in humanitarian settings. To promote strategic investment in technology, leading donors should create a common set of criteria and standards for the responsible deployment of technology in refugee settings, such as drawing on the proposed business model canvas, when determining what projects to fund with NGOs or technology companies.
- *Private companies and implementing partner NGOs* should use this tool when considering specific technologies and use the results of the analysis both to determine which technology products are sensible investments and to improve design and rollout of planned technologies in refugee settings. These stakeholders can use this tool to visualize and fine-tune organizational contributions to the larger ecosystem, maximize humanitarian contributions, and derive value for businesses.

Invest in Sustained and Mainstream Platforms, Data Standards, and Digital Infrastructure

A key finding from this study is that refugees and aid agencies rely mainly on mainstream technology rather than systems or software created expressly for refugees. Yet a tremendous amount of investment and creative energy has been funneled into developing a fragmented and unmaintained set of apps specifically for refugee settings, as described in Chapter Three. Given this, we suggest the following:

- *Aid agencies* should catalogue and prioritize their technology platform, software, and system needs and communicate these to their funders.
- *Host country governments* should identify and invest in specific ways to improve digital infrastructure in their countries for their general populations, which will also improve such systems for refugees.
- *Donors (governments and foundations) and private companies* should focus investments on internet and mobile connectivity and expanding access to mainstream, common platforms for aid agencies, including software, hardware, training, and maintenance. This would make use of government, philanthropic, and private-sector investment more efficient and effective. Instead of investing heavily in pilots and viewing technology platforms as overhead costs ineligible for funding, these stakeholders should provide multiyear funding for sustained technology platforms for aid agencies. The strategy should also include development of common data standards to facilitate interoperability and data exchanges between different platforms. Furthermore, donors should invest in improving digital infrastructure for public services (such as those that include both citizens and refugees) in host countries writ large—for example, by investing in the ten countries hosting the most refugees—in accordance with international norms.

Plan for Technology Scale and Phaseout

Some technical solutions will succeed, while others may not be relevant far into the future. As a result, we suggest the following:

- *Aid agencies and private companies* should develop processes and procedures to periodically assess the continued success and relevance of technology solutions, identify ones that can scale further and across regions and populations, set criteria and procedures for phasing out solutions that have less impact, and strategize about resources that should be reallocated.

Invest in Internet Connectivity, Not New Apps, for Refugees

When refugees had access to internet connectivity and other technology, they made good use of that technology, relying on mainstream platforms. However, refugees have inconsistent access. As a result, we suggest the following:

- *Donors (governments and foundations) and private companies* should focus invest-ments on connectivity for refugees. Specifically, they should include internet con-nectivity in humanitarian aid packages for newly displaced people (because this enables them to keep in contact with family and access other tools necessary to help them in displacement) and invest in Wi-Fi in displacement camps. In addi-tion, these stakeholders should invest in special apps created for refugee settings only when unique circumstances require it.

Improve the Strategic Organization of the Technology Ecosystem Through a Wedding Registry Approach

Investments in technology by donors and aid agencies are fragmented and do not optimally draw on both available solutions or organizations interested and willing to commit resources. Furthermore, multiple consortia with overlapping mandates have formed to assist in contributing to technology in refugee settings, as described in Chap-ter Two. These circumstances lead to fragmented, disconnected investments in one-off solutions rather than systems and more-strategic investments. In particular, organiza-tions that focus on the development of a small-scale app and a narrow population will see less impact than if their efforts had been channeled in a more strategic way in the context of broader strategic deployment of technology.

Instead, many opportunities to partner and contribute effectively may become more apparent through the lens of a systematic framework addressing the entire tech-nology life cycle. Technology needs for refugee settings should be broken into smaller, well-defined projects over a period of years, and stakeholders should solicit field-driven innovation to provide solutions. Each solution should align to the strengths of a single organization, combining to deliver value in a way that is viable in the long term, is financially stable, and contains risks. As a result, we suggest the following:

- *A foundation or a UN agency* should first build the larger vision and plan for technology needed in specific circumstances. By breaking the plan into smaller projects, these can be advertised to donors and technology companies, which can then choose what to fund, understand the long-term value of their contribution, and help estimate risks. Such a wedding registry approach should align the efforts of donors, NGOs, and private companies; reduce risks; and increase the long-term viability of deployed solutions while minimizing duplicated or fragmented resources and risks. The lead foundation or UN agency should coordinate solu-tions to any crucial missing link while allowing for field-driven innovation to fill in the gaps efficiently and effectively.
- *Donors and aid agencies* can help develop and coordinate this wedding registry approach and then align their strengths with this plan, sign on to participate, and execute these smaller projects effectively and efficiently with lower risk.

- *Private companies* should contribute according to the needs identified with this strategic approach. We heard from several interviewees that private companies are eager to contribute through corporate social responsibility commitments but often do not know how. Engaging in strategic investments will help companies optimize their participation, learn from best practices, and focus contributions for impact. It will also help them tailor their business plans and market offerings to the needs of refugees and aid agencies while realizing positive business results. In addition, these companies should develop specifically for aid agencies packages of their technology at discounted prices or with sustainable funding models.

Improve Technical Capacity in the Humanitarian Community

Planning, deploying, using, and maintaining technical solutions require technical, digital transformation, and business process reengineering skill sets. As described in Chapter Three, commonly perceived challenges to effective use of technology in humanitarian settings are the insufficient skill base in aid agencies, the lack of access to needed technological tools in aid agencies, insufficient understanding of the challenges associated with digital transformation, and the need for technology training for refugees. As humanitarian work becomes ever more dependent on digital technology, improving human and institutional technological capacity will be key. As a result, we suggest the following:

- *Private companies* should make further contributions to humanitarian capacity. This should involve creating sustainably discounted software or hardware packages, along with training for aid agencies. The companies should offer additional digital literacy training in refugee camps. They should pay for or provide training for aid agencies; loan staff to aid agencies; and donate consulting services, especially on the topics of digital transformation, change management, and business process reengineering.
- *Aid agencies* should prioritize the use of technology systems and platforms in improving efficiency and invest in hiring and training staff for appropriate technology skills. When accepting volunteered or donated time, aid agencies should design roles that can contribute appropriately to short-term initiatives. Furthermore, they should train and use host country nationals wherever possible to support technology solutions. Doing so can not only increase sustainability but also improve general capacity in the host country workforce.
- *Donor governments and foundations* should pay for professional development in technology for aid agency staff, as well as host country staff working with refugees.
- *Consortia* should continue to draw on the resources of members to support training, platforms, and donations of time and skills.

Improve Effectiveness and Security in Data Management

Humanitarian operations collect, create, use, share, and store vast amounts of data about refugees, aid providers, and operations. Refugees' private data are collected by multiple organizations multiple times, which wastes resources and places unnecessary stress and burden on these vulnerable populations. The policies and procedures for storing, accessing, and securing data are inconsistent at best. Refugees have little understanding of how their data are used and who has access, and they have little recourse for reporting and correcting errors in the data. At the same time, aid agencies have incompatible systems and data formats that make sharing and consolidating data difficult and costly. As a result, we suggest the following:

- *UNHCR* should develop data guidelines for refugee settings that can apply across all aid agencies. These should include policies and procedures to avoid duplication of collection (to the extent possible), security and privacy protections, transparency of use, and refugees' right to access their personal data and report and correct any errors. It should also include policies for expiring unnecessary and obsolete data and procedures for their secure erasure. Based on these guidelines, UNHCR, along with other aid agencies, should develop and implement regional data management plans, to the extent possible and in accordance with national laws and regulations. Finally, aid agencies using refugees' data should periodically conduct risk analyses, balancing the benefits of retaining these vast data sets with the risks of securing the data and preventing misuse and unintended or malicious data breaches.
- *Foundations and private companies* should provide consulting on how UNHCR and other aid agencies could best achieve these recommendations, as well as provide funding and facilitation for such a consultative process in coordination with UNHCR.

Develop an Ethical Framework for Technology in Humanitarian Settings

The growing prevalence of technology in refugee settings has raised multiple unanswered questions about how to use technology in ways that are ethical while balancing effectiveness, efficiency, security, and privacy considerations. Although there are multiple related ethical frameworks, there is no widely accepted ethical framework or set of principles for using digital technology in refugee settings. These issues are further complicated by the complex landscape of the many stakeholders involved and their roles, incentives, and resources. As a result, we suggest the following:

- *UNHCR* should develop an ethical framework for technology in refugee settings, with a set of top-level principles and concrete guidance on applying the principles in various settings. The agency should develop guidelines for evaluating the balance among risks and benefits in using new technologies in refugee settings.

Through our review of the literature and interviews, we found that many organizations aim to introduce value through technology, but they do not have ways of assessing such risk and benefit trade-offs. With such a set of ethical principles and a framework for evaluating risks and benefits, stakeholders may be better able to weigh and then mitigate risks at each stage of the technology life cycle. Such risks and benefits should include human rights, security, privacy, cost, and efficiency concerns, as well as clear descriptions of who benefits and who bears the risks and the alternative approaches available. Over time, there should be a way to discuss and adjudicate ethical issues regarding technology in refugee settings, such as through an entity housed at the UN or affiliated with the United Nations Innovation Forum, with a mechanism to resolve disputes and complaints.

- *A donor government, foundation, or private company* should fund, convene, and facilitate the development of these ethical principles and a risk management framework through a consensus-building process in coordination with UNHCR.

Develop Legal Frameworks Governing Technology, Digital Identity, and Financial Access in Humanitarian Settings in Host Countries

In many cases, host country laws and policies have not caught up to uses of technology, leading to either unregulated or prohibited uses of technology in refugee settings. One of the important findings of this study is that there are gaps in the ability of refugees to present identification documentation needed to access goods and services. Although there are some nascent uses of digital identity for refugees, there remain gaps in both the technology and the policy frameworks that govern and enable their use. Furthermore, refugees often lack access to common digital tools to save and transfer money, make online payments, or take loans. This is often related to host country laws regarding who can access banking services or what types of identification are required to do so. Given that refugees often spend decades in a host country, having access to financial tools is important for their livelihoods and the economies of their host countries. As a result, we suggest the following:

- *Host countries* should carefully consider and develop policies regarding technology in refugee settings, particularly related to data use, digital identities, and financial access. For instance, host countries should develop policies to enable refugees to have proof of identity that enables them to access goods and services, especially banking services and internet or mobile services.
- *UNHCR, in collaboration with leading donors, foundations, and private companies,* should facilitate policy and technical solutions to the need for digital identities and financial access for refugees.

Develop an Improved Evidence Base for Technology in Refugee Education

Educational tools are one of the main ways that private-sector companies have aimed to contribute to refugee situations, yet the evidence base for such tools' effectiveness is thin. Given the growing need for solutions to education for refugees, as well as the private-sector interest in contributing, an improved evidence base is needed so that these investments are not wasted. As a result, we suggest the following:

- *Donors* should fund assessments of effective educational technology in refugee settings, focusing especially on ways that allow refugee education to be scaled up even in contexts with teacher shortages.

Looking Ahead

Through this study, we have found that there is a solid foundation of technology use in humanitarian settings serving a wide variety of needs, and multiple actors create a wide variety of solutions. What is often lacking is the ability to effectively deploy and scale solutions and maintain them over the long run. Fragmented and uncoordinated efforts lead to inefficiencies and do not allow for solutions to be reused across different populations and problem spaces. Part of the problem is a lack of strategic planning and system-level thinking for developing and deploying technology solutions. Furthermore, there is a lack of fundamental understanding of digital transformation and its application to refugees and the humanitarian problem space at large. Too much focus has been placed on the technologies themselves and not enough on how people, culture, and processes should adapt and change to effectively reap the benefits that technology can bring. The economics of markets in refugee settings are not adequately understood—even less so with regard to technology. Future research can shed some light on these topics and guide donors, aid agencies, private companies, and NGOs to collectively provide better services and more access with fewer resources. And although technology will not solve the refugee crisis or even address its underlying fundamental causes, it is improving the lives and livelihoods of refugees worldwide and can do so to a greater extent in the future.

References

"#1425 Hamdi Ulukaya," *Forbes*, 2019. As of August 12, 2019:
https://www.forbes.com/profile/hamdi-ulukaya/#675481861405

AbuJarour, Safa'a, Hanna Krasnova, and Farnia Hoffmeier, "ICT as an Enabler: Understanding the Role of Online Communication in the Social Inclusion of Syrian Refugees in Germany," *European Conference on Information Systems 2018 Proceedings*, 2018.

ActivityInfo, homepage, undated. As of August 14, 2019:
https://www.activityinfo.org/

Agence Française de Développement, "AMAL: Bringing Hope Through Jobs for Young Jordanians and Syrian Refugees," January 31, 2018. As of August 14, 2019:
https://www.afd.fr/en/amal-bringing-hope-through-jobs-young-jordanians-and-syrian-refugees

Airbnb Open Homes, "Host Newcomers Who Are Moving to Your City," undated.

Almohamed, Asam, and Dhaval Vyas, "Designing for the Marginalized: A Step Towards Understanding the Lives of Refugees and Asylum Seekers," *Proceedings of the 2016 ACM Conference Companion Publication on Designing Interactive Systems*, 2016, pp. 165–168.

Associated Press and Louise Boyle, "Chobani Founder Signs Up to Bill Gates' Giving Pledge and Promises to Donate Most of His Self-Made $1.4bn Fortune," *Daily Mail*, May 30, 2015.

Asylum Information Database, "Registration Under Temporary Protection: Turkey," webpage, undated. As of August 14, 2019:
https://www.asylumineurope.org/reports/country/turkey/
registration-under-temporary-protection#footnote3_olljd43

Bacishoga, Kasky Bisimwa, and Kevin Allan Johnston, "Impact of Mobile Phones on Integration: The Case of Refugees in South Africa," *Journal of Community Informatics,* Vol. 9, No. 4, 2013, pp. 1–12.

Baker, Jock, *Humanitarian Capacity-Building and Collaboration: Lessons from the Emergency Capacity Building Project*, London: Humanitarian Practice Network, Network Paper No. 78, June 2014.

Balakrishnan, Anita, "U.N. Turns to Eye-Scanning Technology to Aid Syrian Refugees," NBC News, 2015.

Bansak, Kirk, Jeremy Ferwerda, Jens Hainmueller, Andrea Dillon, Dominik Hangartner, Duncan Lawrence, and Jeremy Weinstein, "Improving Refugee Integration Through Data-Driven Algorithmic Assignment," *Science,* Vol. 359, No. 6373, January 2018, pp. 325–329.

Benton, Meghan, "Digital Litter: The Downside of Using Technology to Help Refugees," Migration Policy Institute, June 20, 2019.

Betts, Alexander, and Louise Bloom, *Humanitarian Innovation: The State of the Art*, New York: United Nations Office for the Coordination of Humanitarian Affairs, November 2014.

Blank, Steve, "The Mission Model Canvas: An Adapted Business Model Canvas for Mission-Driven Organizations," *HuffPost*, December 6, 2017.

Boehm, Barry W., *Software Engineering Economics*, Upper Saddle River, N.J.: Prentice Hall, 1981.

"Brisbane to Help Asylum Seekers and Refugees Resettle in Australia via Technology," *Anthill*, February 16, 2017.

Brynjolfsson, Erik, and Lorin M. Hitt, "Beyond the Productivity Paradox: Computers Are the Catalyst for Bigger Changes," *Communications of the ACM*, Vol. 41, No. 8, 1998, pp. 49–55.

Butcher, Mike, "Here Are 25 of the Most Innovative New Projects Using Tech to Help Refugees and NGOs," *TechCrunch*, October 27, 2018.

Campo, Stuart R., Caitlin N. Howarth, Nathaniel A. Raymond, and Daniel P. Scarnecchia, *The Signal Code: Ethical Obligations for Humanitarian Information Activities*, Cambridge, Mass.: Harvard Humanitarian Initiative, 2018.

Capgemini Consulting, *Technological Innovation for Humanitarian Aid and Assistance*, Brussels: European Parliamentary Research Service, PE 634-411, May 2019.

Carlson, Sam, *Using Technology to Deliver Educational Services to Children and Youth in Environments Affected by Crisis and/or Conflict: Final Report*, Washington, D.C.: U.S. Agency for International Development, December 2013.

Center for Information Technology Research in the Interest of Society and the Banatao Institute, "About," webpage, undated. As of August 13, 2019:
https://citris-uc.org/about/

CHS Alliance, Group URD, and the Sphere Project, *Core Humanitarian Standard on Quality and Accountability*, 2014.

Coppi, Giulio, and Larissa Fast, *Blockchain and Distributed Ledger Technologies in the Humanitarian Sector*, London: Humanitarian Policy Group, February 2019.

Corbett, Jennie, Corinna Frey, and Sonja Marjanovic, "How Innovation Can Assist the Refugee 'Pathway,'" *RAND Blog*, June 27, 2017. As of August 12, 2019:
https://www.rand.org/blog/2017/06/how-innovation-can-assist-the-refugee-pathway.html

Culbertson, Shelly, and Louay Constant, *Education of Syrian Refugee Children: Managing the Crisis in Turkey, Lebanon, and Jordan.*, Santa Monica, Calif.: RAND Corporation, RR-859-CMEPP, 2015. As of August 12, 2019:
https://www.rand.org/pubs/research_reports/RR859.html

Culbertson, Shelly, Olga Oliker, Ben Baruch, and Ilana Blum, *Rethinking Coordination of Services to Refugees in Urban Areas: Managing the Crisis in Jordan and Lebanon*, Santa Monica, Calif.: RAND Corporation, RR-1485-DOS, 2016. As of August 14, 2019:
https://www.rand.org/pubs/research_reports/RR1485.html

Dahya, Negin, *Education in Conflict and Crisis: How Can Technology Make a Difference? A Landscape Review*, Bonn, Germany: Deutsche Gesellschaft für Internationale Zusammenarbeit, 2016.

de Leeuw, Herman, and Stig Arne Skjerven, "Data Mobility in the Fourth Industrial Revolution Age," *University World News*, December 1, 2017.

Dette, Rahel, and Julia Steets, "Innovating for Access: The Role of Technology in Monitoring Aid in Highly Insecure Environments," *Humanitarian Exchange*, No. 66, April 2016, pp. 13–17.

Digital Investment Principles, "The Principles of Donor Alignment for Digital Health," webpage, undated. As of August 15, 2019:
https://digitalinvestmentprinciples.org/

Dimarogonas, James, "A Business Model Canvas for Government Purchases of Commercial Satellite Communications," paper presented at the 30th AIAA International Communications Satellite System Conference, Ottawa, Canada, September 24–27, 2012, pp. 1–16.

Displacement Tracking Matrix, homepage, undated. As of August 14, 2019: https://www.globaldtm.info/

Donahue, Michelle, "This App Is Helping Syrian Refugees Learn to Read," *PCMag*, October 8, 2018.

Duolingo, "About Us," webpage, undated. As of August 13, 2019: https://www.duolingo.com/info

European Parliament, "Asylum: Deal to Update EU Fingerprinting Database," press release, June 19, 2018.

European Union External Action, "About the Global Tech Panel," webpage, September 21, 2018. As of August 13, 2019: https://eeas.europa.eu/headquarters/headquarters-homepage/50886/about-global-tech-panel_en

Field Ready, "Syria," webpage, undated. As of August 14, 2019: https://www.fieldready.org/syria

Fletcher, Tom, *United Networks: Can Technology Help the UN Meet the Challenges of the 21st Century?* New York: Emirates Diplomatic Academy, September 2017.

Frey, Corinna, and Marian Gatzweiler, "How Tech Can Bring Dignity to Refugees in Humanitarian Crises," *The Conversation*, May 2, 2018.

Fuchs, Ben, "The Refugee Crisis: How Coders, Apps, and Technology Provide Relief," *Hackernoon*, March 31, 2016.

Gaffey, Conor, "Five Ways Technology Is Helping with the Refugee Crisis," *Newsweek*, September 28, 2015.

Global Business Coalition for Education, "REACT Initiative," webpage, undated. As of August 13, 2019: https://gbc-education.org/initiatives/gbc-educations-react-initiative/

Goldstein-Rodriguez, Rachel, "UNHCR Seeks ProGres in Refugee Registration," United Nations High Commissioner for Refugees, September 1, 2004.

Google.org, "To Help Those Affected by Crises, We Look to Rapidly Support and Scale Effective Solutions," webpage, undated. As of August 12, 2019: https://www.google.org/our-work/crisis-response/

Gray, Ian, Catherine Komuhangi, Dan McClure, and Lydia Tanner, *Business Models for Innovators Working in Crisis Response and Resilience Building: Exploring Scalable Business Models for Humanitarian Innovation*, Start Network, 2019.

Hempel, Jessi, "How Refugees Are Helping Create Blockchain's Brand New World," *Wired*, March 14, 2018.

Holder, Sarah, "The Algorithm That Can Resettle Refugees," *CityLab*, February 9, 2018.

Hounsell, Benjamin, and Jared Owuor, *Innovating Mobile Solutions for Refugees in East Africa: Opportunities and Barrier to Using Mobile Technology and the Internet in Kakuma Refugee Camp and Nakivale Refugee Settlement*, London: Humanitarian Innovation Fund and Samuel Hall, January 2018. As of August 13, 2019: https://www.elrha.org/wp-content/uploads/2018/02/Innovating_mobile_soultions_Report.pdf

Howell, Catherine, and Darrell M. West, "The Internet as a Human Right," Brookings Institution, November 7, 2016.

Humanitarian Data Exchange, "Frequently Asked Questions," webpage, undated. As of August 14, 2019:
https://data.humdata.org/faq

ID2020 Alliance, "Alliance & Governance," webpage, undated. As of August 12, 2019:
https://id2020.org/partnership/

Indrajit, Sneha, "The Cybersecurity Risks of Using Biometric Data to Issue Refugee Aid," University of Washington, July 25, 2017.

Institute of Electrical and Electronics Engineers, *Standard Glossary of Software Engineering Terminology*, Los Alamitos, Calif.: Computer Society Press, 1990.

Inter-Agency Network for Education in Emergencies, "Technology Task Team," webpage, undated. As of August 13, 2019:
https://inee.org/task-teams/technology

International Federation of Red Cross and Red Crescent Societies and International Committee of the Red Cross, *Code of Conduct for the International Red Cross and Red Crescent Movement and Non-Governmental Organizations (NGOs) in Disaster Relief*, Geneva, undated. As of August 15, 2019:
https://www.icrc.org/en/doc/assets/files/publications/icrc-002-1067.pdf

James, Eric, and Laura James, "3D Printing Humanitarian Supplies in the Field," *Humanitarian Exchange*, No. 66, April 2016, pp. 43–45.

Johnson, Mark W., Clayton M. Christensen, and Henning Kagermann, "Reinventing Your Business Model," *Harvard Business Review,* Vol. 86, No. 12, 2008, pp. 57–68.

Juskalian, Russ, "Inside the Jordan Refugee Camp That Runs on Blockchain," *MIT Technology Review*, April 12, 2018.

Kaplan, Josiah, and Evan Easton-Calabria, "Military Actors and Humanitarian Innovation: Questions, Risks, and Opportunities," *Humanitarian Exchange*, No. 66, April 2016, pp. 26–30.

Kenna, Siobhan, "How Blockchain Technology Is Helping Syrian Refugees," *HuffPost*, November 28, 2017.

Kennedy, Traver, "Energy's 'Last Mile,'" Joi Scientific blog, March 21, 2017.

Khalaf, Roula, "Technology Comes to the Rescue in Migrant Crisis," *Financial Times*, February 24, 2016.

Khan, Shahyan, *Leadership in the Digital Age: A Study on the Effects of Digitalization on Top Management Leadership*, thesis, Stockholm: Stockholm Business School, 2017.

Kuner, Christopher, and Massimo Marelli, eds., *Handbook on Data Protection in Humanitarian Action*, Geneva: International Committee of the Red Cross, 2017.

Lapham, Mary Ann, *Sustaining Software-Intensive Systems: A Conundrum*, Pittsburgh, Pa.: Carnegie Mellon Software Engineering Institute, 2006.

Lapointe, Cara, and Lara Fishbane, "The Blockchain Ethical Design Framework," *Innovations: Technology, Governance, Globalization,* Vol. 12, No. 3–4, 2019, pp. 50–71.

Latonero, Mark, Keith Hiatt, Antonella Napolitano, Giulia Clericetti, and Melanie Penagos, *Digital Identity in the Migration and Refugee Context: Italy Case Study*, New York: Data & Society Research Institute, 2019.

Latonero, Mark, and Paula Kift, "On Digital Passages and Borders: Refugees and the New Infrastructure for Movement and Control," *Social Media + Society,* Vol. 4, No. 1, 2018, pp. 1–11.

Levine, Simon, *Markets in Crisis: The Implications for Humanitarian Action*, London: Humanitarian Policy Group, August 2017.

Lodinová, Anna, "Application of Biometrics as a Means of Refugee Registration: Focusing on UNHCR's Strategy," *Development, Environment and Foresight,* Vol. 2, No. 2, 2016, pp. 91–100.

Maganza, Florian, "Standing with Refugees and Nonprofits That Serve Them on World Refugee Day," Google.org blog, June 20, 2017.

Mason, Ben, "The Failings of Refugee Tech," Betterplace Lab, July 5, 2018.

Mastercard, "Smart Communities Coalition," webpage, undated. As of August 15, 2019: https://www.mastercard.us/en-us/governments/find-solutions/smart-communities.html

Mengel, Sydney, "7 Ways Tech Can Help Refugees," TBD*, January 1, 2018.

Mercy Corps, "A Revolution in Trust: Distributed Ledger Technology in Relief & Development," May 2017.

Migrants' Files, homepage, undated. As of August 14, 2019: https://www.themigrantsfiles.com/

Mikal, Jude P., and Braden Woodfield, "Refugees, Post-Migration Stress, and Internet Use: A Qualitative Analysis of Intercultural Adjustment and Internet Use Among Iraqi and Sudanese Refugees to the United States," *Qualitative Health Research,* Vol. 25, No. 10, 2015, pp. 1319–1333.

Milanio, Leonardo, "#IDETECT: How Technology and Collaboration Between Innovators Can Help Ensure 'No One Is Left Behind,'" Internal Displacement Monitoring Centre, January 2017.

MIT Media Lab, "Refugee Learning Accelerator," webpage, undated. As of August 13, 2019: https://www.media.mit.edu/projects/refugee-learning-accelerator/overview/

Molnar, Petra, and Lex Gill, *Bots at the Gate: A Human Rights Analysis of Automated Decision-Making in Canada's Immigration and Refugee System*, Toronto: International Human Rights Program and the Citizen Lab, 2018.

NetHope, "Our Mission," webpage, undated. As of August 12, 2019: https://nethope.org/our-mission/

NetHope Solutions Center, "Assessing Your Digital Nonprofit Skills," webpage, undated. As of September 30, 2019: https://solutionscenter.nethope.org/the-digital-nonprofit-skills-assessment

Nonnecke, Brandie, "Risks of Recognition," New America blog, September 5, 2017.

Nuttall, Ben, "UNICEF Pi Project to Educate Syrian Children in Lebanon," Raspberry Pi blog, 2014.

Orange Business Services, "13 Ways Humanitarian Organizations Use Tech to Save Lives," February 21, 2018.

Orav, Anita, *Recast Eurodac Regulation*, Brussels: European Parliamentary Research Service, PE 589.808, 2017.

Orcutt, Mike, "How Blockchain Is Kickstarting the Financial Lives of Refugees," *MIT Technology Review*, September 5, 2017.

Osoba, Osonde A., "Keeping Artificial Intelligence Accountable to Humans," *RAND Blog*, August 20, 2018. As of August 15, 2019:
https://www.rand.org/blog/2018/08/keeping-artificial-intelligence-accountable-to-humans.html

Osterwalder, Alexander, and Yves Pigneur, *Business Model Generation: A Handbook for Visionaries, Game Changers, and Challengers*, Hoboken, N.J.: John Wiley & Sons, 2010.

Otieno, Juliet, "#Shupavu291 Success Story: Silas Okinyi from Homa Bay," Eneza Education, April 26, 2017.

"Over 10,000 Migrant Children Missing: Europol," *Hürriyet Daily News*, January 31, 2016.

PeaceTech Lab, homepage, undated. As of August 13, 2019:
https://www.peacetechlab.org/

Pearcy, Aimee, "How We Can Leverage Technology to Help Refugees Gain Employment," *Hackernoon*, May 4, 2018.

Pirlot de Corbion, Alexandrine, Gus Hosein, Tom Fisher, Ed Geraghty, Ailidh Callander, and Tina Bouffet, "The Humanitarian Metadata Problem: 'Doing No Harm' in the Digital Era," Privacy International and International Committee of the Red Cross, October 2018.

Porter, Michael E., "How Competitive Forces Shape Strategy," *Harvard Business Review,* Vol. 57, No. 2, 1979, pp. 137–145.

Principles for Digital Development, homepage, undated. As of August 15, 2019:
https://digitalprinciples.org/

PwC, *Managing the Refugee and Migrant Crisis: The Role of Governments, Private Sector and Technology*, London, 2017.

Radford, Jynnah, and Phillip Connor, "Canada Now Leads the World in Refugee Resettlement, Surpassing the U.S.," Pew Research Center, June 19, 2019.

Rahman, Zara, "Irresponsible Data? The Risks of Registering the Rohingya," *IRIN News*, 2017.

RAND Corporation, "Artificial Intelligence," webpage, undated. As of August 15, 2019:
https://www.rand.org/topics/artificial-intelligence.html

Raymond, Nathaniel A., "Beyond 'Do No Harm' and Individual Consent: Reckoning with the Emerging Ethical Challenges of Civil Society's Use of Data," in Linnet Taylor, Luciano Floridi, and Bart van der Sloot, eds., *Group Privacy: New Challenges of Data Technologies*, Dordrecht, Netherlands: Springer, 2017, pp. 67–82.

Raymond, Nathaniel, Ziad Al Achkar, Stefaan Verhulst, and Jos Berens, *Building Data Responsibly into Humanitarian Action*, New York: United Nations Office for the Coordination of Humanitarian Affairs, May 2016. As of August 13, 2019:
http://datacollaboratives.org/static/files/framework.pdf

Raymond, Nathaniel A., and Casey S. Harrity, "Addressing the 'Doctine Gap': Professionalising the Use of Information Communication Technologies in Humanitairan Action," *Humanitarian Exchange*, No. 66, April 2016, pp. 10–13.

Refugee Open Ware, "National Syrian Project for Prosthetic Limbs, Syria & Turkey," webpage, undated. As of August 14, 2019:
https://row3d.org/initiative/national-syrian-project-for-prosthetic-limbs-syria-turkey

Refugees Welcome International, homepage, undated. As of August 13, 2019:
https://www.refugees-welcome.net/

Ritter, Thomas, and Christopher Lettl, "The Wider Implications of Business-Model Research," *Long Range Planning,* Vol. 51, No. 1, February 2018, pp. 1–8.

Ruhil, Anirudh, "Millions of Refugees Could Benefit from Big Data—But We're Not Using It," *Mashable,* February 21, 2018.

Rumie Initiative, "Learn Syria: Rebuilding Through Education," webpage, 2016. As of August 14, 2019:
https://www.rumie.org/learnsyria/

Rutkin, Aviva, "Phoning in Refugee Aid," *New Scientist,* Vol. 230, No. 3069, April 2016, pp. 22–23.

Ryan, Gery W., and H. Russell Bernard, "Techniques to Identify Themes," *Field Methods,* Vol. 15, No. 1, February 2003, pp. 85–109.

Schiemichen, Laura, "Emerging ID Technology Helps Refugees, at a Cost to Privacy," *Kennedy School Review*, March 27, 2018.

Schmidt, Douglas C., "The Growing Importance of Sustaining Software for the DoD: Part 1," Software Engineering Institute blog, August 1, 2011.

Scott, Brett, "How Can Cryptocurrency and Blockchain Technology Play a Role in Building Social Solidarity Finance?" United Nations Research Institute for Social Development Working Paper 2016-1, February 2016.

Scriven, Kim, "Humanitarian Innovation and the Art of the Possible," *Humanitarian Exchange*, No. 66, April 2016, pp. 5–7.

Sebti, Bassam, "4 Smartphone Tools Syrian Refugees Use to Arrive in Europe Safely," *World Bank Blogs*, February 17, 2016.

Simko, Lucy, Ada Lerner, Samia Ibtasam, Franziska Roesner, and Tadayoshi Kohno, "Computer Security and Privacy for Refugees in the United States," paper presented at the 2018 IEEE Symposium on Security and Privacy, May 20–24, 2018.

Skirble, Rosanne, "Shelved Machine Translator Gets New LIfe in Haiti Relief Effort," Voice of America, February 7, 2010.

Soesilo, Denise, and Kristin Bergtora Sandvik, *Drones in Humanitarian Action: A Survey on Perceptions and Applications*, Geneva: Swiss Foundation for Mine Action, 2016.

Soliman, Sarah, "Tracking Refugees with Biometrics: More Questions Than Answers," *War on the Rocks*, March 9, 2016.

Souder, William E., "Improving Productivity Through Technology Push," *Research-Technology Management,* Vol. 32, No. 2, March–April 1989, pp. 19–24.

Spelhaug, Justin, "Accelerating Transformation for Nonprofits with Technology for Social Impact," Microsoft blog, February 21, 2018.

Sphere Association, *The Sphere Handbook 2018: Humanitarian Charter and Minimum Standards in Humanitarian Response*, 4th ed., Geneva, 2018.

Stulman, Michael, "Escaping Boko Haram Attacks in Nigeria," *Medium*, February 23, 2017.

Tali, Didem, "Four AI-Powered Technologies Aimed at Helping Refugees," Dell Technologies, August 14, 2018.

Tarjemly Live, homepage, undated. As of August 13, 2019:
http://tarjemly-live.com/en/

Tausan, Michaelle, and Luke Stannard, *EdTech for Learning in Emergencies and Displaced Settings: A Rigorous Review and Narrative Synthesis*, London: Save the Chidren UK, 2018.

Techfugees, "About Us," webpage, undated. As of August 12, 2019:
https://techfugees.com/about/

Tent Partnership for Refugees, homepage, undated. As of August 12, 2019:
https://www.tent.org/members/

Thaki, homepage, undated. As of August 14, 2019:
http://thaki.org/

TikkTalk, "About Us," undated-a. As of August 13, 2019:
https://www.tikktalk.com/en/about-us/

———, "Customers," undated-b. As of August 13, 2019:
https://www.tikktalk.com/en/customers/

———, "Refugee Camps," undated-c. As of August 13, 2019:
https://www.tikktalk.com/en/category/refugee-camps/

Tomaszewski, Brian, "Teaching Refugees How to Map Their World Could Have Huge Benefits," *Smithsonian*, May 21, 2018.

Tomaszewski, Brian, Jean-Laurent Martin, and Yusuf Hamad, "GIS for Refugees, by Refugees," *ArcNews*, Vol. 39, No. 3, Summer 2017, pp. 4–5.

Toor, Amar, "Syrian Refugees Share Memories Stored on Their Phones in Powerful Photo Series," *The Verge*, March 27, 2017.

Transformify, "Rebuild Lives Program: Hire Refugees and Revitalize Post-War Zones," webpage, undated. As of August 13, 2019:
https://www.transformify.org/page/rebuild-lives-program-hire-refugees-and-revitalize-post-war-zones

Translators Without Borders, "Development & Preparedness," webpage, undated-a. As of August 13, 2019:
https://translatorswithoutborders.org/our-work/development/

———, "Our Work," webpage, undated-b. As of August 13, 2019:
https://translatorswithoutborders.org/our-work/

UNESCO—*See* United Nations Educational, Scientific and Cultural Organization.

UNICEF—*See* United Nations Children's Fund.

UNHCR—*See* United Nations High Commissioner for Refugees.

UNOCHA—*See* United Nations Office for the Coordination of Humanitarian Affairs.

United Nations, "About the Sustainable Development Goals," webpage, undated. As of November 6, 2019:
https://www.un.org/sustainabledevelopment/sustainable-development-goals/

United Nations Children's Fund, "UNICEF Lebanon Showcases Raspberry Pi for Learning," Stories of Innovation blog, August 7, 2014.

———, "Humanitarian Drone Corridor Launched in Malawi," August 23, 2017.

United Nations Department of Global Communications, "Data for Democracy Wins Unite Ideas #IDETECT Data Challenge to Monitor Worldwide Patterns of Internal Displacement," Relief Web, June 22, 2017.

————, "The United Nations System," January 1, 2019. As of August 12, 2019:
https://www.un.org/en/pdfs/18-00159e_un_system_chart_17x11_4c_en_web.pdf

United Nations Educational, Scientific and Cultural Organization, *A Lifeline to Learning: Leveraging Technology to Support Education for Refugees*, Paris, 2018.

United Nations General Assembly, *Guidelines for the Regulation of Computerized Personal Data Files*, Geneva, December 14, 1990.

United Nations Global Compact, "United Nations Private Sector Forum 2016," 2016. As of August 12, 2019:
https://www.unglobalcompact.org/take-action/events/691-united-nations-private-sector-forum-2016

United Nations High Commissioner for Refugees, "Basic Assistance," webpage, undated-a. As of August 14, 2019:
https://www.unhcr.org/lb/basic-assistance

————, "Biometric Identity Management System," undated-b. As of August 14, 2019:
https://www.unhcr.org/en-us/protection/basic/550c304c9/biometric-identity-management-system.html

————, "From proGres to PRIMES," undated-c. As of August 14, 2019:
https://www.unhcr.org/blogs/wp-content/uploads/sites/48/2018/03/2018-03-16-PRIMES-Flyer.pdf

————, "New York Declaration for Refugees and Migrants," webpage, undated-d. As of August 12, 2019:
https://www.unhcr.org/new-york-declaration-for-refugees-and-migrants.html

————, "Non-Governmental Organizations," webpage, undated-e. As of August 12, 2019:
https://www.unhcr.org/en-us/non-governmental-organizations.html

————, "PRIMES," webpage, undated-f. As of August 14, 2019:
https://www.unhcr.org/en-us/primes.html

————, "UNHCR Strategy on Digital Identity and Inclusion," undated-g. As of August 14, 2019:
https://www.unhcr.org/blogs/wp-content/uploads/sites/48/2018/03/2018-02-Digital-Identity_02.pdf

————, "Humanitarian Principles," in *Emergency Handbook*, version 1.7, Geneva, 2015a. As of August 15, 2019:
https://emergency.unhcr.org/entry/223864/humanitarian-principles

————, "Policy on the Protection of Personal Data of Persons of Concern to UNHCR," Geneva, May 2015b.

————, *Global Trends: Forced Displacement in 2015*, Geneva, June 20, 2016a.

————, *Connecting Refugees: How Internet and Mobile Connectivity can Improve Refugee Well-Being and Transform Humanitarian Action*, Geneva, September 2016b.

————, *Missing Out: Refugee Education in Crisis*, Geneva, September 2016c.

————, "Mobile Connectivity a Lifeline for Refugees, Report Finds," September 14, 2016d.

————, *Report of the United Nations High Commissioner for Refugees: Part II—Global Compact on Refugees*, New York, A/73/12 (Part II), 2018.

————, "Donors," webpage, March 21, 2019a. As of August 12, 2019:
https://www.unhcr.org/en-us/donors.html

————, *Global Trends: Forced Displacement in 2018*, Geneva, June 20, 2019b.

United Nations Innovation Network, "About the UN Innovation Network," webpage, undated. As of August 13, 2019:
https://www.uninnovation.network/about-us

———, "Innovation Quarterly Update (Q3)," 2018. As of November 1, 2019:
https://drive.google.com/file/d/1xmcbl2hvSgB3rYYuxJFC_i86uvii5Slv/view

United Nations Office for the Coordination of Humanitarian Affairs, "The Future of Technology in Crisis Response," April 11, 2017.

U.S. Agency for International Development, *Identity in a Digital Age: Infrastructure for Inclusive Development*, Washington, D.C., undated.

USAID—*See* U.S. Agency for International Development.

Vosloo, Steve, "New Evidence on What Educational Technology Works for Refugees and Displaced Populations," ICTworks, April 7, 2018.

Wagner, Emma, *Refugee Education: Is Technology the Solution?* London: Save the Children UK, 2017.

Waughman, Adele, "From Principles to Practice: Implementing the Principles for Digital Development," Washington, D.C.: Principles for Digital Development Working Group, January 2016.

Weiss-Meyer, Amy, "Apps for Refugees: How Technology Helps in a Humanitarian Crisis," *The Atlantic*, May 2017.

Wilding, Raelene, "Mediating Culture in Transnational Spaces: An Example of Young People from Refugee Backgrounds," *Continuum,* Vol. 26, No. 3, 2012, pp. 501–511.

W. K. Kellogg Foundation, *Logic Model Development Guide*, Battle Creek, Mich., January 2004.

World Bank, *Principles on Identification for Sustainable Development: Toward the Digital Age*, Washington, D.C., February 2017.

———, "World Bank Group, UNHCR Sign Memorandum to Establish Joint Data Center on Forced Displacement," April 20, 2018.

Zikusooka, Monica, Zinet Nezir Hassen, Alison Donnelly, and Rachel Mose, "Conducting Simulated Field Visits for Insecure Locations in Somalia," *Humanitarian Exchange*, No. 66, April 2016, pp. 18–20.

Zott, Christoph, Raphael Amit, and Lorenzo Massa, "The Business Model: Theoretical Roots, Recent Developments, and Future Research," Barcelona: University of Navarra IESE Business School Working Paper, WP-862, September 2010.